National Center for Construction Education and Research

AF863748

Rigging

PEARSON
Prentice Hall

Upper Saddle River, New Jersey
Columbus, Ohio

contren®
Learning Series

National Center for Construction Education and Research

President: Don Whyte
Director of Product Development and Revision: Daniele Stacey
Rigging Project Manager: Natalie Smith
Production Manager: Jessica Martin
Product Maintenance Supervisor: Debie Ness
Editor: Bethany Harvey
Desktop Publisher: Jennifer Jacobs

The NCCER would like to acknowledge the contract service provider for this curriculum:
Topaz Publications, Liverpool, New York.

10 9 8 7 6
ISBN 0-13-227296-2

Preface

This volume was developed by the National Center for Construction Education and Research (NCCER) in response to the training needs of the construction, maintenance, and pipeline industries. It is one of many in NCCER's *Contren® Learning Series*. The program, covering training for close to 40 construction and maintenance areas, and including skills assessments, safety training, and management education, was developed over a period of years by industry and education specialists.

NCCER also maintains a National Registry that provides transcripts, certificates, and wallet cards to individuals who have successfully completed modules of NCCER's *Contren® Learning Series*, when the training program is delivered by an NCCER Accredited training Sponsor.

The NCCER is a not-for-profit 501(c)(3) education foundation established in 1995 by the world's largest and most progressive construction companies and national construction associations. It was founded to address the severe workforce shortage facing the industry and to develop a standardized training process and curricula. Today, NCCER is supported by hundreds of leading construction and maintenance companies, manufacturers, and national associations, including the following partnering organizations:

PARTNERING ASSOCIATIONS

- American Fire Sprinkler Association
- American Petroleum Institute
- American Society for Training & Development
- American Welding Society
- Associated Builders & Contractors, Inc.
- Associated General Contractors of America
- Association for Career and Technical Education
- Carolinas AGC, Inc.
- Citizens Development Corps
- Construction Industry Institute
- Construction Users Roundtable
- Design-Build Institute of America
- Electronic Systems Industry Consortium
- Merit Contractors Association of Canada
- Metal Building Manufacturers Association
- National Association of Minority Contractors
- National Association of State Supervisors for Trade and Industrial Education
- National Association of Women in Construction
- National Insulation Association
- National Ready Mixed Concrete Association
- National Systems Contractors Association
- National Utility Contractors Association
- National Technical Honor Society
- North American Crane Bureau
- North American Technician Excellence
- Painting & Decorating Contractors of America
- Portland Cement Association
- SkillsUSA
- Steel Erectors Association of America
- Texas Gulf Coast Chapter ABC
- U.S. Army Corps of Engineers
- University of Florida
- Women Construction Owners & Executives, USA

Some features of NCCER's *Contren® Learning Series* are:

- An industry-proven record of success
- Curricula developed by the industry for the industry
- National standardization providing portability of learned job skills and educational credits
- Credentials for individuals through NCCER's National Registry
- Compliance with Apprenticeship, Training, Employer, and Labor Services (ATELS) requirements for related classroom training (CFR 29:29)
- Well-illustrated, up-to-date, and practical information

Acknowledgments

This curriculum was revised as a result of the farsightedness and leadership of the following sponsors:

Austin Bridge & Road, Inc.
ExxonMobil Baton Rouge
ExxonMobil Baytown
Fluor Enterprises, Inc.
Halliburton
JE/Jacobs
JE Merit Constructors
KBR
Shaw Stone and Webster
Turner Industries Holding Co.
Zachry Construction Corporation

This curriculum would not exist were it not for the dedication and unselfish energy of those volunteers who served on the Authoring Team. A sincere thanks is extended to:

Doyle Adams
Adam Clement
Dale DeLapp
Tony Guidry
Jim Harbor
Paul Kuber
David Olivarez
Samuel Rogers
David Sisk
Steven Slocum
Vernon Smith
Mike Terry
Don Totty
Anthony Townsend
Julio Yglesias, Jr.

Contents

38201-06 Wire Rope 1.i

38202-06 Telescopic Boom Assembly and Disassembly 2.i

38203-06 Lattice Boom Assembly and Disassembly 3.i

38204-06 Lift Planning 4.i

38205-06 Advanced Rigging Practices 5.i

38206-06 Personnel Lifting 6.i

Rigging . . . . .Index

Module 38201-06

Wire Rope

COURSE MAP

This course map shows all of the modules in the second part of the *Rigging* curriculum, *Rigging*. The suggested training order begins at the bottom and proceeds up. Skill levels increase as you advance on the course map. The local Training Program Sponsor may adjust the training order.

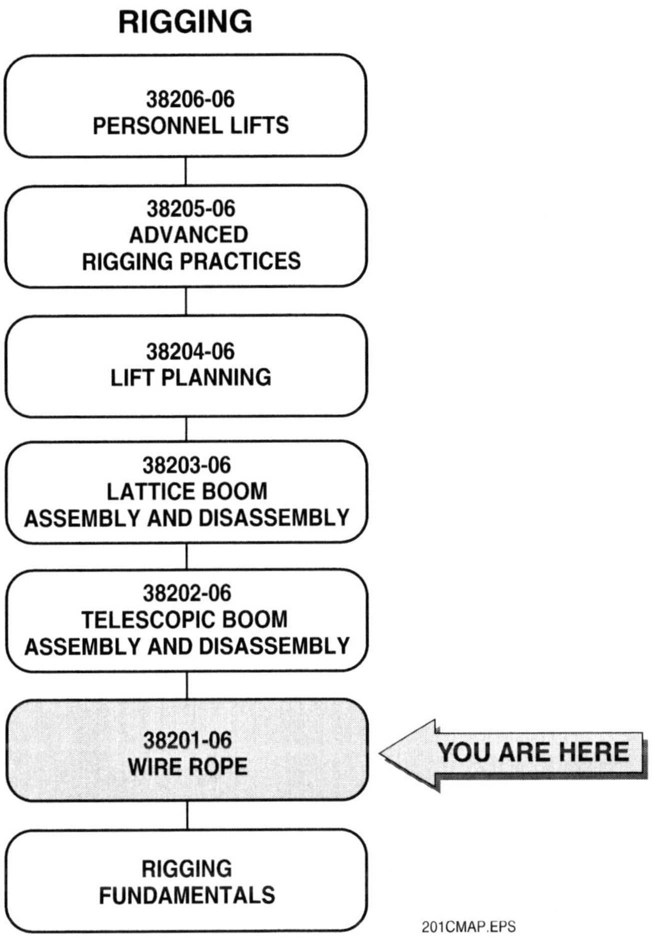

201CMAP.EPS

MODULE 38201-06 CONTENTS

1.0.0 INTRODUCTION 1.1

2.0.0 WIRE ROPE CONSTRUCTION 1.1

2.1.0 Wire Rope Components 1.1

2.2.0 Wire Rope Component Configuration 1.2

2.3.0 Identification of Wire Rope 1.4

2.4.0 End Terminations and Preparations 1.4

2.4.1 Fabricated Socket 1.4

2.4.2 Wedge Socket (Becket Assembly) 1.5

2.4.3 Wire Rope Clips 1.7

3.0.0 INSPECTION AND MAINTENANCE REQUIREMENTS 1.7

3.1.0 Inspection Guidelines 1.7

3.2.0 Inspection Criteria 1.8

3.3.0 Sheave and Load Block Inspection 1.9

3.4.0 D/d Ratio 1.10

3.5.0 Maintenance Guidelines 1.10

4.0.0 WIRE ROPE REEVING 1.11

4.1.0 Determining Parts of Line 1.12

4.2.0 Block Twisting 1.13

5.0.0 REEVING A DRUM 1.13

5.1.0 Fleet Angle 1.14

SUMMARY 1.15

REVIEW QUESTIONS 1.16

GLOSSARY 1.19

ACKNOWLEDGMENTS 1.21

Figures

Figure 1 Components of a wire rope 1.1
Figure 2 Right regular lay wire rope 1.2
Figure 3 Left regular lay wire rope 1.2
Figure 4 Right lang lay wire rope 1.2
Figure 5 Left lang lay wire rope 1.2
Figure 6 Right alternate lay wire rope .1.3
Figure 7 Four basic strand patterns 1.3
Figure 8 Strand rope classifications 1.3
Figure 9 End preparations 1.4
Figure 10 End fittings 1.5
Figure 11 Wedge socket components 1.6
Figure 12 Wedge socket end termination 1.6
Figure 13 Wire rope clips 1.7
Figure 14 Examples of wire rope failure 1.8
Figure 15 Load and hook diameter and D/d ratio chart for wire rope .1.11
Figure 16 Reeving diagrams 1.12
Figure 17 Drum reeving 1.14
Figure 18 Storage reel to drum 1.14
Figure 19 Fleet angle 1.15

Wire Rope

Objectives

When you have completed this module, you will be able to do the following:

1. Describe how wire rope is constructed and secured and how its breaking strength is determined.
2. Determine the allowable working load of wire rope.
3. Perform a wire rope inspection.
4. Identify wire rope replacement criteria and describe procedures for replacement.
5. Describe the proper maintenance procedures for wire rope.
6. Describe proper procedures and methods of reeving all wire ropes and multiple-part lines.
7. Describe the advantages of using multi-part reeving.

Prerequisites

Before you begin this module, it is recommended that you successfully complete *Rigging Fundamentals*.

Required Trainee Materials

1. Pencil and paper
2. Appropriate personal protective equipment

1.0.0 ◆ INTRODUCTION

A mobile crane has enormous power that can be used to lift loads and move them to a desired location. To lift and move a load, the crane must be able to convert its power and provide a means to allow the work to be performed on the load. A crane transmits its power to the hoist **drum** and onto the load through the **wire** rope. Wire rope is a very sophisticated device specifically designed to support lifting operations. This module presents information about the construction, care, inspection, and use of wire rope.

2.0.0 ◆ WIRE ROPE CONSTRUCTION

An understanding of the construction of wire rope and its components is critical in performing proper inspections and giving the rope proper care. There are hundreds of different kinds of wire rope available. This section will provide a broad understanding of the types that are available. This section will also provide a working knowledge of wire rope to enable riggers to use it safely.

2.1.0 Wire Rope Components

A wire rope consists of only three parts. The method of construction and relation of the parts to each other give each specific type of rope its unique characteristics. The three parts are the wire, the **strands**, and the **core**, as shown in *Figure 1*.

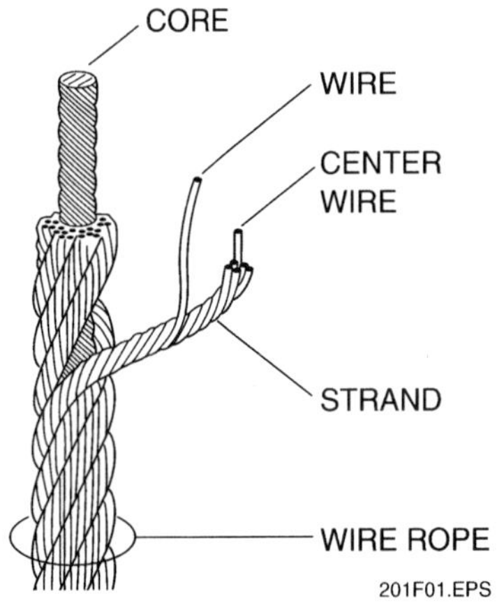

Figure 1 ◆ Components of a wire rope.

The wire can be made of many different materials and can be shaped to many different forms. It can be made of metals such as steel, iron, bronze, and **Monel™**. Because of the wide variety and availability of various **grades** of steel, it is the most common type of material used in the construction of wire rope. The most common grades of wire rope steel, in order of increasing breaking strengths, are as follows:

- *Traction steel (TS)* – Traction steel wire rope is used in elevator service.
- ***Plow steel (PS)*** – Plow steel has lower tensile strength, but it can be used when strength is secondary to wear resistance.
- *Improved plow steel (IPS)* – Improved plow steel is the most commonly used wire rope. It is wear-resistant, has high tensile strength, and is fatigue-resistant.
- *Extra improved plow steel (EIPS)* – This rope is used when the need for higher breaking strength is required. The bending quality of this rope is not as good as that of improved plow steel.
- *Extra extra improved plow steel (EEIPS)* – This rope is used when special installations require maximum rope strength, such as mine-shaft hoisting.

The term plow steel is a reference to an old grading system for steel developed to provide a consistent gauge in determining the tensile strengths of steel. This term is still used today as a reference for grading wire rope.

Strands are made up of two or more wires in a specific pattern. A strand can be assembled from several types of material in addition to the steel wires. Wire ropes can be made of any number of strands fitted together. Typical strand grades are: common, Siemens-Martin, high-strength, and extra-high strength.

The core provides support for strands under normal working conditions. The core can be made of several different materials including steel and synthetic fibers. Typical wire rope core types can be identified as **fiber core (FC), independent wire rope core (IWRC)**, or **wire strand core (WSC)**.

2.2.0 Wire Rope Component Configuration

Wire rope component configuration describes the pattern of the components of wire rope. This is important in identifying a wire rope and its possible use. To understand this aspect of wire rope manufacturing, operators must first understand the relationship between the strands, and of the strands to the core of the rope. This relationship or pattern is known as the lay of the rope. This term addresses the make-up of the direction of wrap of the strand **lay (strand)** and the length of the strand **lay (length)**.

The strand-to-core relationship is addressed by the first term, as either right or left lay. Right lay indicates that the wrap of the strand is in a clockwise direction around the core, as shown in *Figure 2*. Left lay indicates that the direction of wrap is in the counterclockwise direction, as shown in *Figure 3*.

201F02.EPS

Figure 2 ◆ Right regular lay wire rope.

201F03.EPS

Figure 3 ◆ Left regular lay wire rope.

The terms regular, lang, and alternate lay refer to the strand-to-strand relationship. *Figures 2* and *3* are examples of regular lay wire rope. Regular lay indicates that the individual wires of each strand run parallel to the direction of the run of the rope. Therefore, the rope in *Figure 2* is considered a right regular lay rope, while *Figure 3* is a left regular lay rope.

Lang lay is indicated when the individual wires of a strand are perpendicular to the run of rope. *Figure 4* is an example of a right lang lay wire rope and *Figure 5* is an example of a left lang lay wire rope. When the strands of a wire rope are a combination of regular and lang lay, it is referred to as alternate lay. *Figure 6* shows an example of right alternate lay wire rope. The majority of the rope encountered on a crane will be right regular lay wire rope.

201F04.EPS

Figure 4 ◆ Right lang lay wire rope.

201F05.EPS

Figure 5 ◆ Left lang lay wire rope.

Figure 6 ◆ Right alternate lay wire rope.

Classification of a wire rope also involves the number of wires, the pattern in which they are laid, and their geometric pattern. *Figure 7* is an example of the four basic **strand patterns** in which strands are laid: wire, Warrington, Seale, and **filler wire**.

An additional classification of wire rope is based on the number of strands and the number of wires in each strand. These are nominal numbers and do not necessarily reflect the exact number of wires in each strand. As an example of this system, 6 × 7 indicates that there are 6 strands in the rope and 7 wires make up each strand; 6 × 19 would indicate 6 strands and 19 wires per strand (see *Figure 8*). In the case of some wires (for instance, the 6 × 19 class of wires), 6 × 21 filler wire and all 6 × 25 filler wire would both fall into the 6 × 19 class. Each rope contains 6 strands; however, each strand will have more than 19 wires.

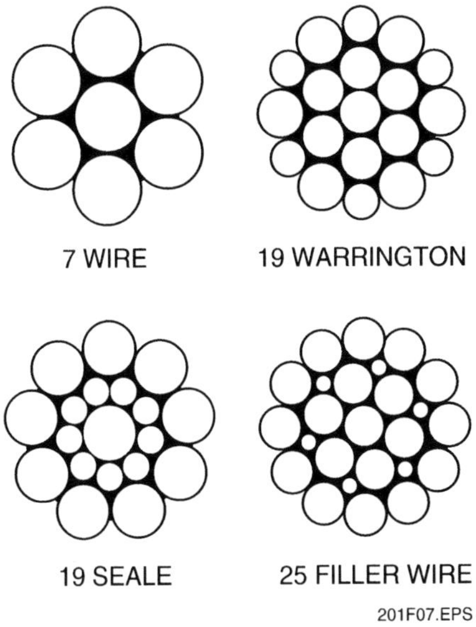

Figure 7 ◆ Four basic strand patterns.

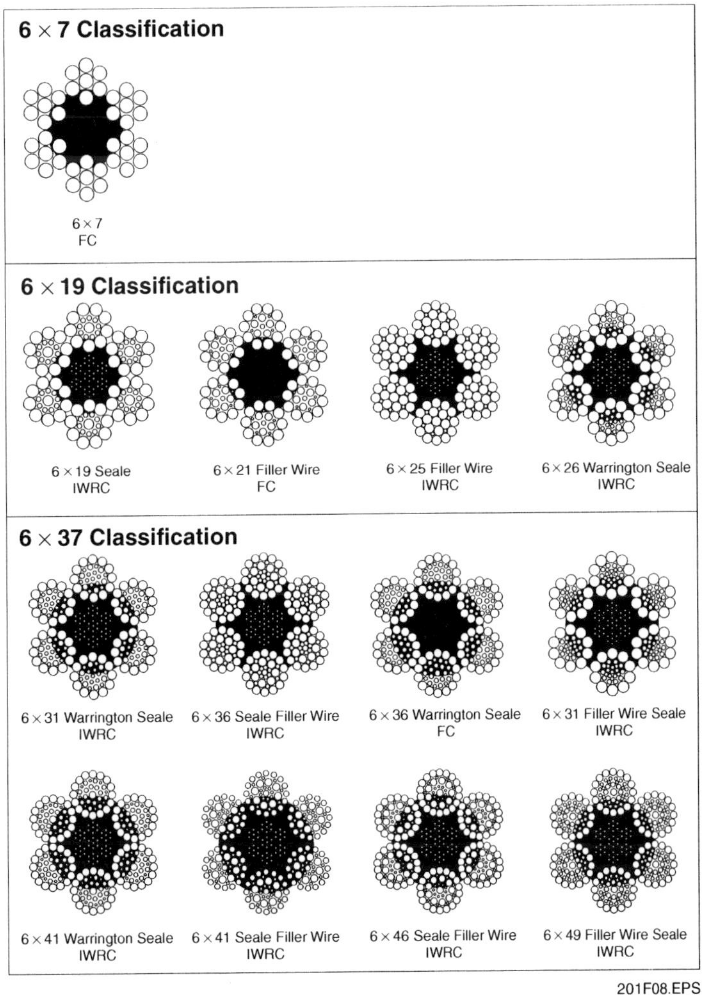

Figure 8 ◆ Strand rope classifications.

Another descriptive characteristic of wire rope is the finish of the wire. Most wire is made from high carbon steel. Unless a special coating is added to change the finish, it will have a bright finish. Some wire ropes are manufactured with special protective coatings such as zinc, resin, or plastic.

2.3.0 Identification of Wire Rope

Now that you have been introduced to the characteristics of wire rope, the next step is to identify it using these characteristics. In the identifying description of wire rope, the following characteristics are used: length, diameter, finish, lay, grade, core type, and whether it is **preformed** (pref) or non-preformed (non-pref).

As an example, a description of wire rope may read as follows:

1,200 ft ¾" 6 × 19 WS pref RRL EIPS IWRC

This description indicates that there are 1,200 feet of ¾" diameter rope. The rope is 6 × 19 Warrington Seale. It consists of 6 strands in a Warrington Seale (WS) combination, and each strand will have 19 wires.

This type of wire is preformed and is a right regular lay (RRL) rope. Each individual wire is made from high carbon steel of the extra improved plow steel (EIPS) grade. The strands are wrapped around an independent wire rope core (IWRC).

2.4.0 End Terminations and Preparations

Both ends of a wire rope must be secured in some manner. On a mobile crane's load hoist rope, one end is normally secured to the drum, and the other is affixed to a point on the **load block** or the boom tip. This enables the power of the machine to be transmitted to the load for lifting purposes.

Occasionally, the wire rope end will require preparation to enable it to be **reeved** or threaded through the **sheaves** and to the connector on the drum. The types of end preparations vary. They can include either a welded and tapered end, or **seizing**, as shown in *Figure 9*. The method will be specified by the manufacturer. After the end preparation is completed, the rope can then be threaded through the narrow opening provided by the sheaves and guards and the drum connection point.

The other end of the wire rope must also be secured, which allows the load block or ball to lift the load without unreeving the rope. As shown in

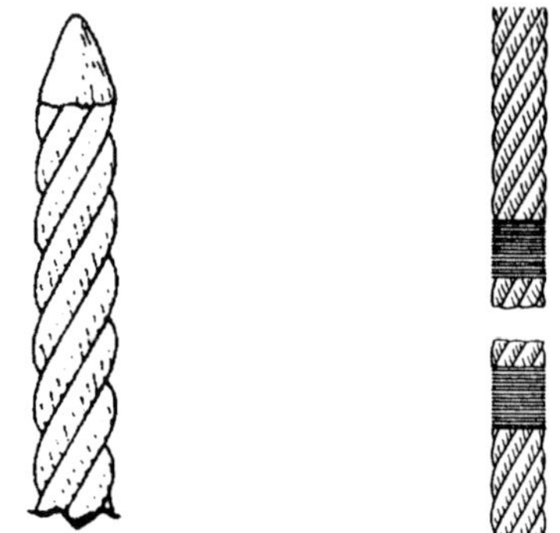

Figure 9 ◆ End preparations.

Figure 10, end fittings can consist of any of the following types:

- Poured socket
- Swaged socket
- Mechanical or hand-tucked splice (**loop** or thimble)
- Wedge socket
- Wire rope clips

As shown in the illustration, each of these end terminations has a specific **efficiency**, expressed as a percentage of the nominal wire rope breaking strength. As an example, if the termination is a poured socket (either resin or **spelter**), which has an efficiency rating of 100 percent, this means the termination is as strong as the wire rope.

A very popular termination, the wedge socket, has an efficiency of 75 to 80 percent, depending on the design. This means that the wedge socket has 75 to 80 percent of the **nominal breaking strength**. This indicates that it is weaker than the wire rope by 20 to 25 percent when compared to the nominal breaking strength of the wire rope. This could possibly be a failure point if all the components of a wire rope assembly were **stressed** to capacity.

2.4.1 Fabricated Socket

There are two types of fabricated sockets: poured and swaged. Poured sockets consist of a metal cup (cone) with an end termination attached. The wire rope is attached to the socket with liquefied zinc, which is poured into the cup through which the wire rope is inserted. When the metal cools and solidifies, it holds the rope firmly in place. The hot metal connection is commonly referred to as a speltered socket. A variation on this type of socket

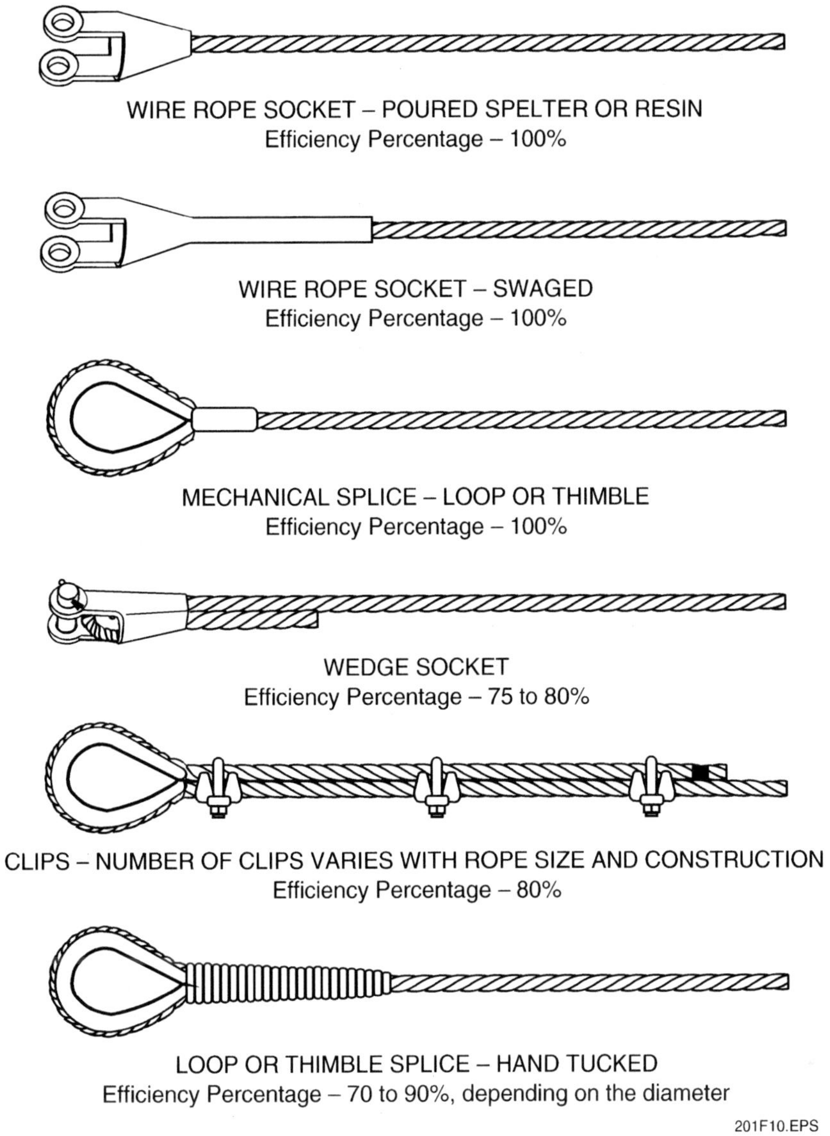

Figure 10 ◆ End fittings.

replaces the hot metal with an epoxy resin, which hardens through a chemical reaction, holding the wire rope in place. The efficiency of both types is 100 percent of nominal rope strength. The selection of which type of socket to use will depend on its application because the resin is vulnerable in a high-heat or chemically caustic atmosphere.

Swaged sockets are visually similar to poured sockets. The most striking difference is in how the wire rope is secured. In the swaged socket, the wire rope is secured mechanically. Swaging is a process in which the base of the fitting (shank) is cold pressed or crimped tightly to the wire rope, forming a mechanical bond. If the fitting is properly pressed and the wire rope properly prepared, this type of end termination has a 100 percent efficiency rating when used on wire rope constructed with an independent wire rope core (IWRC). This type of fitting is not recommended for use when the wire rope is constructed with a fiber core (FC).

2.4.2 Wedge Socket (Becket Assembly)

The wedge socket is a very popular end termination. It consists of a wedge, wedge socket, and pin (*Figure 11*). A wire rope loop is threaded through the wedge socket. A wedge is then placed in the loop and pushed back into the socket. As a load is applied, the wedge will seat tighter into the wedge socket, holding the wire rope in place. The shank on the wedge socket has a removable pin affixed. It is used for mounting the load attachment device, such as a hook, swivel, **headache ball**, or load block.

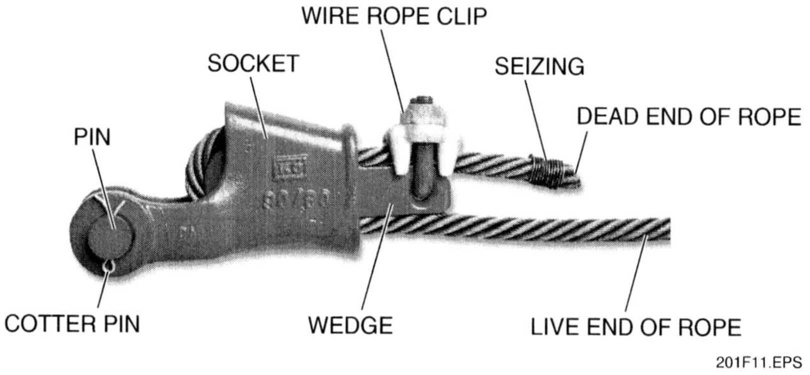

Figure 11 ◆ Wedge socket components.

The socket, wedge, and pin should be inspected for correct size and condition before use. Do not use parts that are corroded, cracked, or worn. The operator should consult the socket manufacturer for recommendations regarding the specific use and reapplication of wedge sockets.

> **NOTE**
> Careful attention should be given to wedge socket components. They are a matched pair and should be kept together.

To understand the end termination, you must know how to determine the **live section** of a wire rope from the **dead end**. This concept is easily explained by the following definitions: the live section of a wire rope is the section that will be under load. The dead end does not experience the weight of the load.

When assembling a wedge socket, there are important checkpoints to keep in mind to prevent incorrect assembly:

- Do not attach the dead end to the live section of the wire rope.
- Secure the dead end of the wire rope.
- Align the live section of the wire rope with the center line of the pin.
- Exercise care when seating the wedge into the socket with a hammer. Do not strike the wire rope.
- The tail length of the dead end of wire rope should be a minimum of six rope diameters but not less than six inches.

Figure 12 shows that there are correct and incorrect methods of positioning the wire rope, as well as the termination of the dead end of the rope. The section of wire rope in line with the pin will experience the load weight; therefore, it is considered the live part of the line. In contrast, the end that wraps around the wedge and exits the socket is considered the dead end, which does not experience the weight of the load. Securing the dead end of the wire rope prevents it from slipping out of the socket in an unloaded condition. Always follow manufacturer's recommendations when installing a wedge socket.

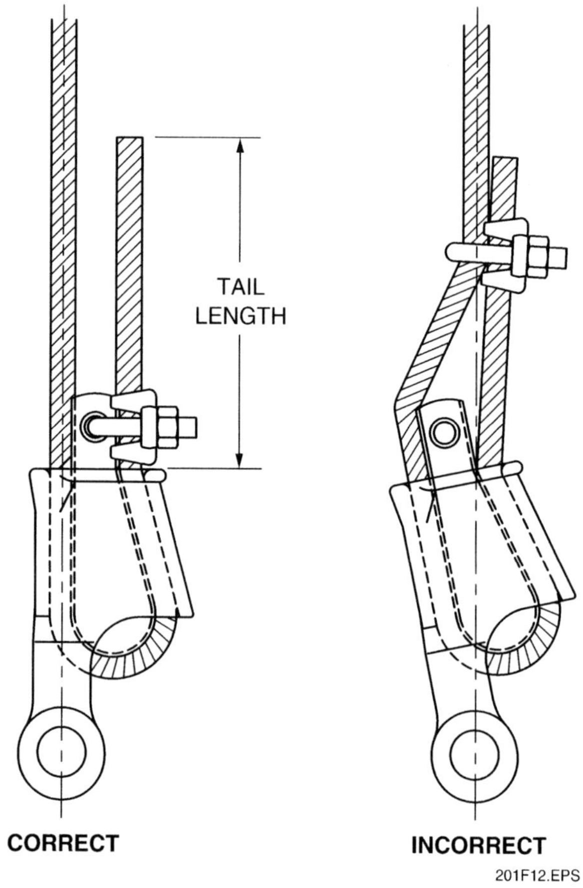

Figure 12 ◆ Wedge socket end termination.

2.4.3 Wire Rope Clips

Wire rope clips are used to create end terminations such as thimbles and to terminate wedge socket assemblies. There are two common designs for wire rope clips: the forged U-bolt and the fist grip (*Figure 13*). The most obvious difference between the two is the location of the mounting nuts. The U-bolt clip has two U-bolt nuts on the same side of the clip, while the fist grip has mounting nuts on opposite sides of the clip.

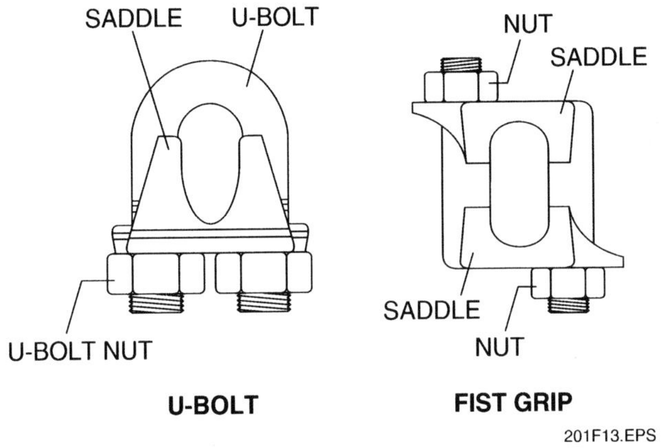

Figure 13 ◆ Wire rope clips.

To properly mount U-bolt wire rope clips, you must first identify the location of the live section and dead end of the rope. The U-bolt clip consists of the U-bolt, the nuts, and the saddle. The direction that the U-bolt and the saddle face is critical to achieving full strength for the application.

In all applications, the saddle is positioned on the live section of wire rope, and the U-bolt contacts the dead end. The design of the saddle provides protection for the live section of rope, preventing crushing and deformation when the U-bolt nuts are tightened during the installation procedure.

The fist grip wire rope clip has the unique advantage of having no right or wrong direction for mounting. The two sides are mirror images of each other, having a saddle on each, which protects the wire rope when the mounting bolts are tightened.

A common practice when creating assemblies using more than one wire rope clip is to space them evenly. The nuts should also be tightened evenly and alternately until they reach the recommended torque. It is mandatory to test load the assembly. The test load weight should be equal to or greater than the weight of the expected load. After testing, the nuts should again be tightened to the specified torque.

> **NOTE**
> To avoid damage, the manufacturer's specifications should always be followed for testing wire rope clips.

3.0.0 ◆ INSPECTION AND MAINTENANCE REQUIREMENTS

A planned and thoroughly documented program of periodic inspection and maintenance is required to meet regulatory requirements and to ensure that the wire rope will be safe for use and provide maximum service life. An experienced inspector must be able to determine the cause of various rope degradations and make judgments on the continued service life of the wire rope based on observations.

3.1.0 Inspection Guidelines

A good inspection program will provide guidelines to inspectors that aid in determining the condition of a wire rope and its suitability for continued service. The following guidelines are based on those established by regulatory agencies, consensus standards, and those compiled by the Wire Rope Technical Board.

- *Inspection records* – Review the available inspection records. Ensure that they are maintained for the required length of time as prescribed by regulations.
- *Frequent inspections* – Frequent inspections are those that meet the following criteria: All running rope in service on mobile cranes should be visually inspected once each working day for gross damage. There is also a statutory requirement to have a written, dated, and signed monthly report on file. A frequent inspection should look for the following:
  - **Kinks**, crushing, **birdcaging**, unstranding, main strand displacement, or core protrusion
  - General **corrosion**
  - Broken or cut strands and wires
  - Core failure

> **NOTE**
> Loss of rope diameter in a short length or unevenness of the outer strand should be evidence that the rope is an immediate hazard and should be removed from service immediately.

- *Periodic inspections* – Periodic inspections should be conducted by an authorized competent inspector and meet the requirements of the applicable standard. There are statutory requirements that mandate that certain items be checked annually and that a signed report be completed. The items that are checked on a periodic inspection of the wire rope and reeving include the following:
  - All items listed in the frequent inspection section
  - Reduction of rope diameter below nominal diameter due to loss of core support, internal or external corrosion, or wear of outside wires
  - Severely corroded or broken wires at the end connections
  - Severely corroded, cracked, bent, worn, or improperly applied end connections
  - Sections in contact with saddles, equalizer sheaves, or other sheaves where rope travel is limited
  - Sections of the rope at or near terminal ends where corroded or broken wires may protrude
  - Wire rope slings are checked for excessive wear, broken wires, kinking, twisting, and mechanical abuse
  - All end connections such as hooks, shackles, turnbuckles, plate clamps, and sockets are checked for excessive wear

3.2.0 Inspection Criteria

When inspecting wire rope, an understanding of the following criteria is important and is the foundation for becoming a knowledgeable inspector. Some examples of wire rope failure are shown in *Figure 14*.

- ***Abrasion*** – Abrasion occurs when wire rope moves through an abrasive medium or over drums and sheaves. Most standards require rope to be removed from service if the wear of the outer wires exceeds one-third of the original outer diameter.
- *Rope* ***stretch*** – All rope will initially stretch as load is applied. Wire rope stretches in three phases. Phase 1 is the initial phase, in which a rope will stretch as its initial load is applied (due to setting of the wires and strands to each other); this is known as constructional stretch. Phase 2 is the most prolonged period of stretch. This phase of stretch will see only small changes in rope diameter due to the normal wear and **fatigue** of the rope. Phase 3 is the end of life stretch; this is a point of rapid deterioration of the rope and of significant changes in rope diameter.

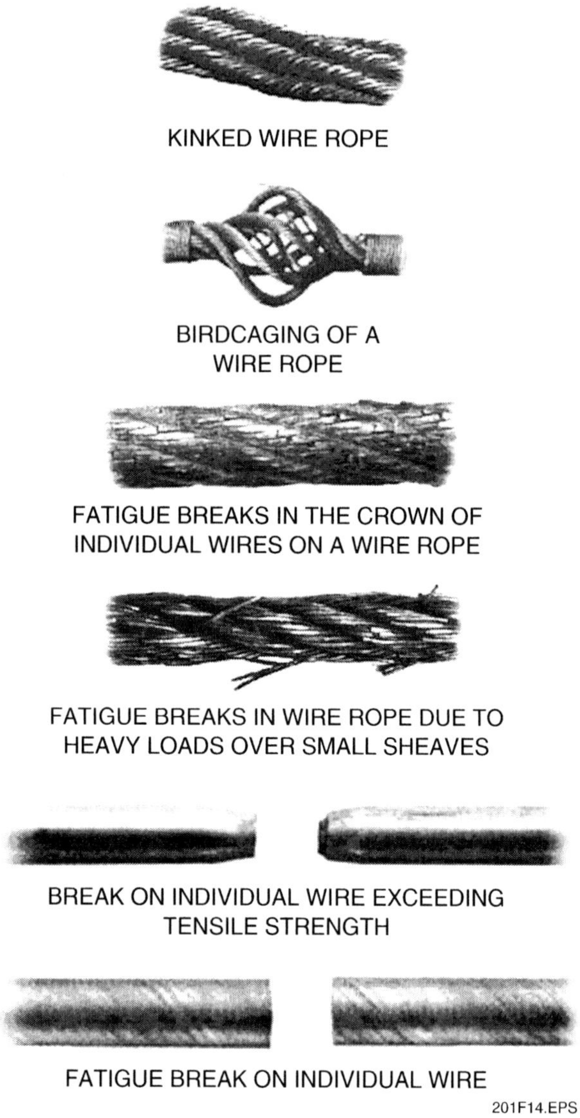

Figure 14 ◆ Examples of wire rope failure.

- *Reduction in rope diameter* – Any marked reduction in rope diameter is a strong indication of degradation. This can occur at the end of a wire rope's service life. This can also be an indication of excessive external abrasion, internal or external corrosion, loosening or tightening of rope lay, **inner wire** breakage, and rope stretch of strands.
- *Corrosion* – Corrosion is a more serious cause of degradation than abrasion and is usually indicative of a lack of lubrication. Since wire rope corrosion usually occurs internally before there is any external indication, pitting of the wires is grounds for immediate removal from service. Pitting is due to the degradation of the internal wires and their inability to move smoothly over each other during operation.
- *Kinks* – Kinks are permanent distortions caused by loops drawn too tightly. Ropes with kinks must be removed from service.

- *Birdcaging* – Birdcaging is caused by mistreatment of the wire rope, such as sudden stops, the rope being pulled through tight sheaves, or the rope wound on too small a drum. This is cause for rope replacement or removal of that section of wire rope.
- *Localized condition or abrasion* – These localized conditions or abrasions can occur at equalizing sheaves. During normal operations, this wear may not be visible. Excessive vibration or whip can cause abrasion or fatigue. Drum crossover and flange point areas must also be examined closely. All end fittings, including splices, should be examined for worn or broken wires, loose or damaged strands, cracked fittings, worn or distorted thimbles, and tucks of strands.
- *Heat damage* – High temperatures may cause metal discoloration or loss of internal lubrication; fiber core ropes are particularly vulnerable.
- *Electric arc* – Electric arc occurs when the wire rope contacts an energized power source or is used as a ground for an electric welding unit. As electricity passes through the wire rope, some of the wires will be fused, **annealed**, and discolored, requiring replacement.
- *Protruding core* – Core protrusion occurs from an opening between the strands. This is cause for removal from service.
- *Damaged end attachments* – End fittings that are cracked, bent, or broken must be eliminated. The cause should be determined and corrected. In the case of bent hooks, the throat openings measured at the narrowest point should not exceed 15 percent over normal; there should not be twisting greater than 10 degrees.
- *Peening* – One cause of peening (metal-to-metal denting and deformation) is from the wire rope continuously striking a structural element of the crane, a roller, or the wire rope itself. Another cause is continuous work under high load over a sheave or drum. If this situation cannot be controlled, the rope must be inspected more often and could require more frequent replacement.
- *Scrubbing* – Scrubbing is the displacement of wires and strands caused by rubbing. This causes accelerated wear and displacement of wires and strands along one side of the rope.
- *Fatigue fracture* – Fatigue fractures are best identified at the breaking point, which will have square ends and show little sign of surface wear. This type of failure usually occurs at the crown of the strands or in the valleys between strands where adjacent strand contact exists. In most cases, these failures are related to bending or vibration in sheaved or reeved areas of the wire rope.
- *Broken wires* – The number of broken wires in a wire rope is an indication of the general condition of the wire rope. Using this as a guide, some criteria have been established to mandate wire rope replacement based on the number of wires in the rope that are broken. Once broken wires start to appear in a wire rope, it is prudent to assume that many more will appear in a short time under normal operating conditions.

The guidance provided by the industry consensus standards suggests that wire rope operating on steel drums and sheaves needs to be replaced if any of the following broken wire conditions exists:

- Six randomly distributed broken wires in one rope lay (length) or three broken wires in one strand of one wire rope lay
- Two randomly broken wires in an area of six rope diameters or four randomly broken wires in thirty rope diameters (requirement for rotation resistant rope)
- One broken outer wire at the point of contact with the core that physically protrudes or loops out of the wire rope structure
- Standing rope with more than two broken wires in one lay in sections beyond the end connections or more than one broken wire at an end connection

3.3.0 Sheave and Load Block Inspection

The sheave and load block, which work in conjunction with the hook and headache ball, are an integral part of the wire rope system. They must be inspected as precisely as wire ropes. Most mobile cranes require the wire rope to contact one or more sheaves and a load block when attached. Because of this wire rope contact, the physical condition of the load block and the sheaves can have a tremendous effect on the service life of wire rope. Integrating the inspection of crane sheaves and the load block into the wire rope inspection program can have long-term benefits.

NOTE

The load block contains markings showing its load and capacity. The correct load block must be used in accordance with the manufacturer's specifications.

The focus of the information presented in this section will be on the sheaves mounted on the crane and on the load block sheaves. For this discussion, the two critical areas that will be addressed are the bearings and groove dimensions.

A sheave is designed to support and guide the wire rope. A load block consists of a series of sheaves stacked in a support frame with a hook attached. When examining load blocks and crane sheaves, check the bearings and their lubrication. Sheaves should turn easily, not wobble, and be lubricated in accordance with the manufacturer's documentation. Sheaves that are not properly lubricated will increase the amount of energy necessary to move a load due to an increased friction. This added stress can shorten the service life of the wire rope. Bearings that are worn out from improper maintenance or normal wear can result in a wobble of the sheave. This wobble can cause significant vibration of the wire rope, resulting in accelerated wear. Another area of the sheaves to check is the condition of the groove and the flanges. The flanges must be checked for physical damage that may snag the wire rope or cause it to jump or walk out of the groove when it is rotating. This type of physical damage can also result in cutting, curling, burring, or crowning of the wire rope strands.

Metal-to-metal contact can be very abrasive, as is the contact between the wire rope and the sheave groove. Under normal conditions, the wire rope will wear the groove on a sheave. As the groove wears, it provides less support for the wire rope, allowing it to flatten out and move in the groove. This added movement increases the speed at which the wire rope and sheaves deteriorate. Another way that a sheave can wear a wire rope is by compressing or pinching the wire rope in the groove. For example, the sheave is worn in such a manner that the groove narrows. In this case, the wire rope passes through the narrower opening of the worn sheave, pinching the strand together. This action increases the internal friction of the rope, accelerating its wear.

3.4.0 D/d Ratio

The D/d ratio (*Figure 15*) is the relationship between the pitch diameter of the sheave and the rope diameter. This factor is important in establishing the fatigue resistance or relative service life of the wire rope. When the tread D/d ratio is too small, it can increase the rope and sheave wear significantly. This wear can be directly attributed to the increased bending stress applied to the rope and sheave when the wire rope is forced into too tight a bend.

Manufacturers calculate minimum ratios for sheaves, drums, and wire ropes. By comparing the minimum ratio with the calculated ratio, it can be determined if the different products are compatible with each other for safety and durability. As an example, it is desired to use a brand X ½-inch-diameter wire rope on a brand Y boom hoist sheave. The manufacturer of the wire rope has a suggested minimum D/d ratio of 30. To determine the minimum size sheave that must be mounted on the boom, multiply the ½-inch nominal rope diameter by its minimum suggested D/d ratio of 30. The result is 15 inches, which is the minimum diameter sheave that should be mounted on the boom to maintain the minimum D/d ratio for the ½-inch wire rope. This comparison should be made each time a sheave, drum, or wire rope is replaced with material other than that recommended by the manufacturer.

3.5.0 Maintenance Guidelines

A short list of acceptable maintenance practices for wire rope is supplied in this section. This will provide you with a good foundation on which to base a maintenance program when working with wire rope. Any long-range inspection and maintenance program should include a good records program. Keeping accurately dated records can assist in predicting the service life for a wire rope. It also helps in determining labor and cost requirements for job planning.

- Wire rope should always be stored in a manner that prevents damage or deterioration.
- Always follow the manufacturer's recommendations when removing wire rope from its reel or spool.
- Always avoid kinking or twisting the wire rope.
- To prevent unlaying of the strands, place seizing on wire rope before cutting.
- To prevent scraping, nicks, crushing, or exceeding its bend radius, do not drag the wire rope through dirt or around sharp objects.
- Lubricate the wire rope on a regular basis, as recommended by the manufacturer. This reduces internal friction and prevents corrosion. The lubricant must be consistent with that originally used by the manufacturer and should not hinder visual inspection of the wire rope.
- Consideration should be given to shortening a wire rope to move a section that shows slight signs of increased wear to an area of lesser wear. This prevents running the same section of rope over the same sheave, spreading the wear over more of the rope and thus extending its service life.

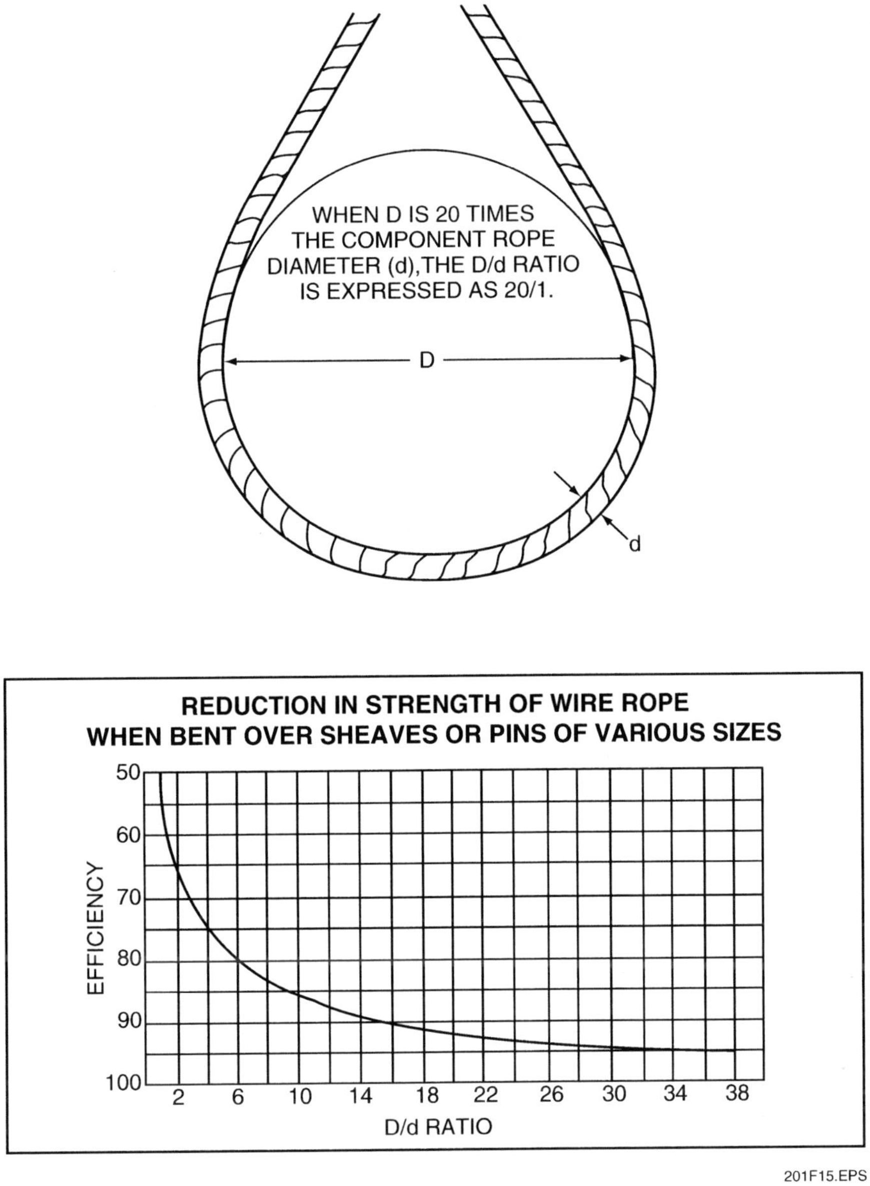

Figure 15 ◆ Load and hook diameter and D/d ratio chart for wire rope.

4.0.0 ◆ WIRE ROPE REEVING

The most common way to use wire rope attached to a mobile crane is to reeve on a load block. Attaching the load block is simply placing the wire rope in a specified manner around the sheaves. This placing of the wire rope in the sheaves is known as reeving (*Figure 16*). The number of times the wire rope passes through the block, leaving a span of wire from one sheave to the next sheave, is known as a part of line. As more parts of line are added to the load block, the amount of weight pulling on the wire rope at any given point or single line pull is reduced. This reduction of single line pull can be attributed to the fact that the total weight of the load will be equally shared between the parts of line reeved on the load block.

CAUTION

To avoid cable damage or failure, wire rope must be reeved per manufacturers' requirements according to load chart configurations.

WARNING!

Replacement wire rope is to have a strength rating at least as great as the original rope furnished or recommended by the manufacturer. Any change from the original size, grade, or construction is to be specified by a rope manufacturer, the crane manufacturer, or a qualified person.

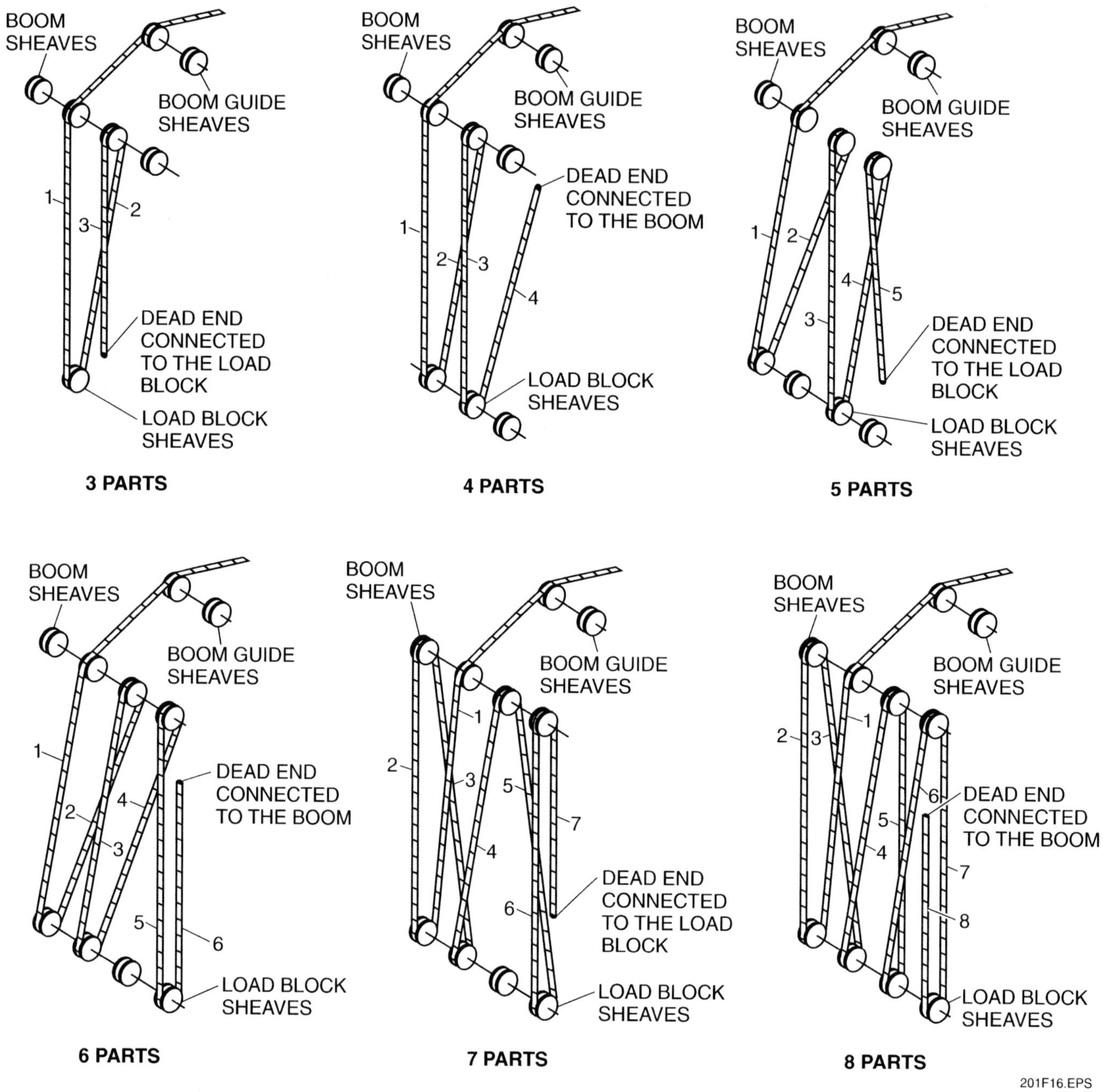

Figure 16 ◆ Reeving diagrams.

4.1.0 Determining Parts of Line

Several methods are used to determine the parts of line for a lift. Some crane manufacturers place this information in the load charts. This is the easiest of the methods. This method is also accurate if all material used is that recommended by the manufacturer.

Another method is a formula and chart provided by equipment suppliers that can be used to determine the number of parts of line. This method uses a ratio calculated by dividing the total weight of the load to be lifted (to be safe, this is everything suspended below the boom tip) by the single line pull, which is the catalog nominal breaking strength of the wire rope. This ratio is then matched to a table that has required parts of line listed. This table also includes a consideration for the type of bearing or bushing that the sheave uses. The bearing and bushing consideration is necessary to account for frictional loads associated with the sheave.

The last method is to manually calculate the parts of line using the following formulas:

- Nominal breaking strength (catalog breaking strength as determined by the manufacturer) divided by the **design factor** (safety factor) equals the allowable load for the wire rope (allowed single line pull). The safety factors are as follows:

- For supporting loads, live or running rope is 3.5 and boom pendant or standing rope is 3.0.
- For boom erection purposes, live or running rope is 3.0 and boom pendant or standing rope is 2.5.
- For rotation-resistant rope, the factor is 5.0.

- Total load (everything below the boom tip), divided by the single line pull (safe working load) equals the number of parts of line required for the lift. Anything less than a whole number is rounded up to the nearest whole number.

In this method, a consideration should be made to account for the load friction of the sheaves.

4.2.0 Block Twisting

Block twisting is frequently encountered when using a load block. This problem is directly associated with the construction of wire rope. When the load is placed on the wire rope, it has a tendency to unlay or remove the twist inserted during the manufacturing process. As the load increases, the amount of unlaying will increase. This untwisting or unlaying will rotate the load block. The rotation of the load block can cause severe problems during a lift, up to and including binding of the load block and sheaves.

Several actions can be taken to minimize the amount of block twisting that occurs. First, reduce the wire rope length. Long wire rope is more susceptible to block twisting than shorter ropes. Second, reduce the load. Block twisting increases at higher loads. Reducing the load and the amount of shock loading decreases block twisting.

Another way to reduce block twisting is to remove the odd part of line by reeving the block to an even number of parts. The dead end of the rope should be terminated at the boom to increase its separation from the other parts of line. Another action to increase the separation of the parts of line is to use larger sheaves; this reduces the amount of torque applied to the block.

An additional method is to use a tag line attached to the load. The block must be locked. This provides a stabilizing counter-torque to the block.

Rotation-resistant rope is used in some applications with the manufacturer's approval, but it does not guarantee that the rope won't twist.

5.0.0 ◆ REEVING A DRUM

The component that actually transfers the power of the crane to the wire rope is the hoist drum. The drum provides the physical connection of the wire rope to the crane. The hoist drum holds or moves the load up and down as required for the job. Since the interface between the crane and the load is critical to the safe operation of the crane, the method for reeving the wire onto the drum is also critical.

Two types of drums can be encountered on a crane: grooved and smooth. A grooved drum has channels in which the first layer of wire rope is wound. The second and subsequent layers of wire rope lay in the valleys formed by the coils of the first layer. It is imperative that the drum be reeved correctly. If the wire rope is carelessly wound and jumps the grooves, it will be crushed and cut where it crosses from one groove to another.

A smooth drum has no grooves to guide the first layer. However, just as with the grooved drum, the subsequent layers of the smooth drum lay in the valleys of the previous layer. Correct reeving is equally essential with a smooth drum. Loose and uneven winding on a plain-faced (smooth) drum can create excessive wear and crushing and distortion of the rope. This results in low operating performance, a reduction in the rope's effective strength, and in some instances, breaking of the rope.

Wire rope drums are reeved using one of four directional methods (*Figure 17*). To determine the proper direction to wind the wire rope, stand behind the crane and point down the wire rope path with the index finger. Use the hand that is in the direction of the lay of the rope; for example, use the left hand for left lay and the right hand for right lay wire rope.

If the wire rope is to be wound over the top (overwind), point with the palm of the hand facing down. If the wire rope is wound in the opposite direction (underwind), have the palm of the hand facing up. Using this procedure, the direction to apply left or right lay rope on a smooth drum can be easily determined.

One last critical element for reeving wire rope is its proper removal from the storage reel to the drum. The preferred direction is from the top of the reel to the top of the drum (*Figure 18*) or from the bottom of the reel to the bottom of the drum. This prevents placing a reverse bend in the wire rope, causing it to become twisty and hard to handle.

NOTE

As required by OSHA, there must be a minimum of two wraps remaining on the drum when the load block is at its lowest point.

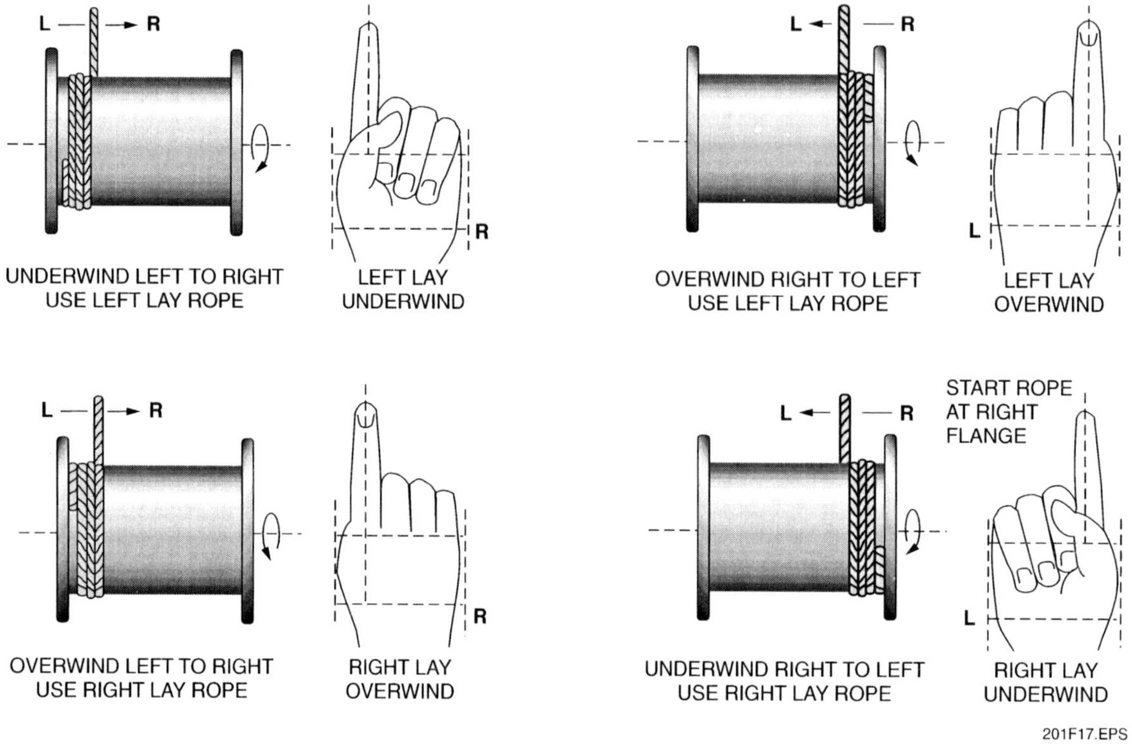

Figure 17 ◆ Drum reeving.

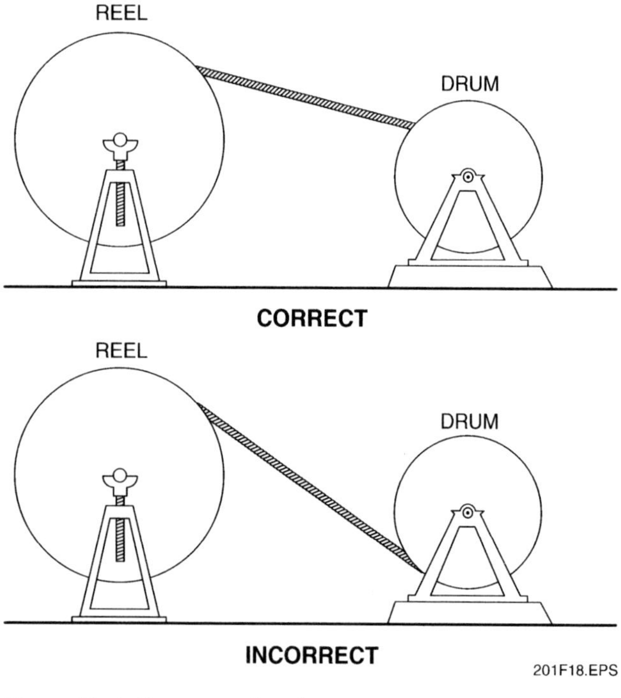

Figure 18 ◆ Storage reel to drum.

5.1.0 Fleet Angle

When a wire rope is run over a sheave and onto a drum, the rope will not be in direct alignment with the groove of the sheave. The amount of deviation, which is called the fleet angle (*Figure 19*) depends on the width of the drum and the distance from the sheave.

It is desirable to keep the fleet angle at no more than 1½ degrees for smooth drums and 2 degrees for grooved drums. This angle can be controlled by adjusting the distance of the lead sheave from the drum (DLS).

The standards for distance are 38 feet per foot of drum width for smooth drums and 29 feet per foot of drum width for grooved drums. For example, assume a grooved drum that is 3-feet wide from flange to flange. Half that distance (DCF) is 1.5 feet. $29 \times 1.5 = 43.5$. Therefore, a DLS of 43.5 feet is needed to maintain the desired fleet angle.

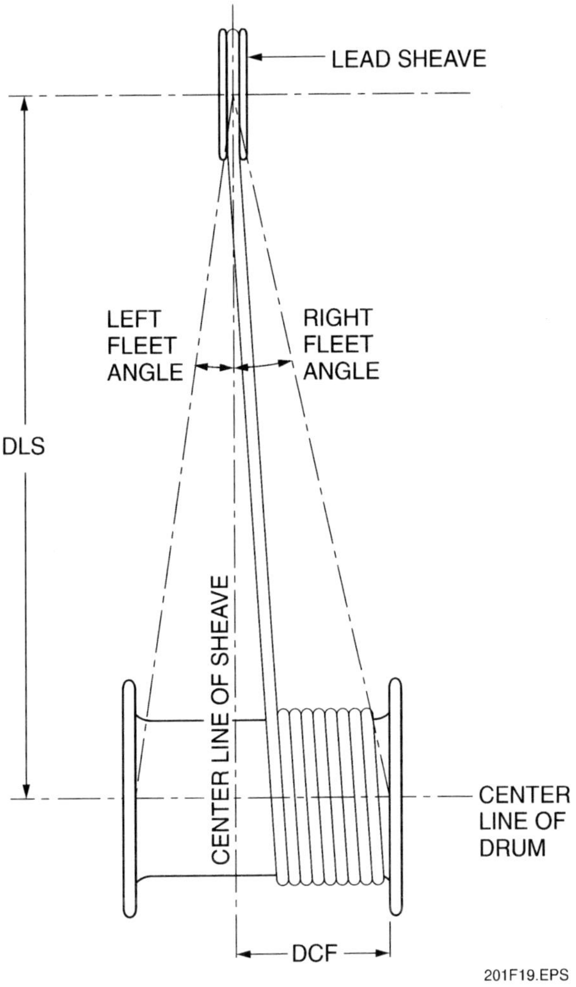

Figure 19 ◆ Fleet angle.

Summary

One of the critical links from the crane to the load is through the wire rope. It allows the power of the crane to be transmitted to the load, resulting in lift and movement. As a rigging specialist, your knowledge and understanding of wire rope construction, use, and maintenance practices are critical to the safe operation of the crane. An essential part of any routine inspection program for cranes should include inspecting the wire rope. This module has provided a good foundation on which to start building in-depth knowledge required for performing wire rope inspections.

Review Questions

1. The three parts of a wire rope are the wire, the strands, and the _____.
   a. Monel™
   b. center wire
   c. core
   d. lang

2. The grade of steel with the highest breaking strength is _____ steel.
   a. traction
   b. plow
   c. improved plow
   d. extra extra improved plow

3. The wire used in rope can be made of steel, iron, Monel™, or _____.
   a. copper
   b. aluminum
   c. bronze
   d. tin

4. Wire rope cores are identified as fiber core, wire strand core, or _____ core.
   a. filler
   b. independent wire rope
   c. rotation resistant
   d. Seale

5. What type of lay is indicated when the strands are wrapped around the core in a counter-clockwise direction?
   a. Right lay
   b. Left lay
   c. Alternate lay
   d. Lang lay

6. When the individual wires of a strand on a rope are perpendicular to the run of the rope it is classified as _____ lay.
   a. regular
   b. left
   c. alternate
   d. lang

7. The four basic strand patterns are wire, Warrington, filler, and _____.
   a. warrel
   b. contractor
   c. Seale
   d. laquor

8. Within the 6 × 37 classification of rope, the _____ can vary from 31 to 49.
   a. number of wires in a strand
   b. number of strands
   c. diameter of the core
   d. pattern of the strands

Match the rope identification with the characteristics in 9 and 10.

9. _____ The longest

10. _____ The thickest

   a. 1,100 ft ¾ inch 6 × 19 WS pref RRL EEIP IWRC
   b. 1,300 ft ½ inch 6 × 19 WS pref RLL IPS FC
   c. 1,400 ft ¼ inch 6 × 37 WS pref LLL EIPS IWRC
   d. 1,200 ft ¼ inch 6 × 37 WS pref RRL EIPS IWRC

11. All rope ends must be welded and tapered to prevent fraying.
    a. True
    b. False

12. The lowest efficiency rating of a wire rope with a loop or thimble splice, hand tucked, is _____ percent.
    a. 70
    b. 80
    c. 90
    d. 95

13. The efficiency rating of a wire rope assembly is a percentage of the _____.
    a. nominal breaking strength
    b. abrasion ratio
    c. applied load
    d. parts of a line

14. What type of end termination provides the highest efficiency rating?
    a. Thimble, hand tucked
    b. Swaged socket
    c. Wedge socket
    d. Wire rope clips

15. A fabricated socket has an efficiency rating of _____ percent.
    a. 70 to 95
    b. 75 to 80
    c. 80
    d. 100

16. A swaged socket is not recommended for _____ type of wire rope.
    a. IWRC
    b. FC
    c. WSC
    d. EIPS

17. The dead end of the wire rope experiences the weight of the load.
    a. True
    b. False

18. In a wedge socket, the tail length of the dead end of the rope should be at least _____.
    a. long enough for three rope clips
    b. six centimeters
    c. six rope diameters
    d. ten inches

19. The U-bolt type of clip consists of the U-bolt, the nuts, and the _____.
    a. socket
    b. clip ring
    c. saddle
    d. clip harness

20. How often should the running rope of a mobile crane be inspected for gross damage?
    a. Daily
    b. Weekly
    c. Monthly
    d. Annually

21. As part of the inspection criteria, most standards require rope to be removed from service if the wear on the outer wire exceeds _____ of the original diameter.
    a. one-eighth
    b. one-fifth
    c. one-third
    d. one-half

22. Some minor defects in the wire rope can be tolerated, but if any _____ is/are found during inspection, the wire rope must be immediately removed from use.
    a. heat damage
    b. broken wires
    c. peening
    d. kinks

23. A wire rope needs to be replaced if there is/are _____ broken outer wire(s) at the point of contact with the core that physically protrudes or loops out of the wire structure.
    a. one
    b. three
    c. four
    d. six

24. Damaged flanges on a sheave can cause the wire rope to _____.
    a. wobble
    b. stretch
    c. birdcage
    d. jump out of the groove

25. The two critical areas to inspect on a sheave are the bearings and the _____.
    a. safety latch
    b. rope terminations
    c. groove dimensions
    d. gauge

26. A series of sheaves stacked in a support frame with a hook attached is a _____.
    a. sheave stack
    b. load block
    c. boom sheave
    d. boom pendant

27. The D/d ratio is used to _____.
    a. calculate the length of rope needed
    b. check if different products are compatible
    c. calculate the number of parts of a line needed
    d. measure the efficiency of a wire rope termination

28. To reduce internal friction and prevent corrosion, a wire rope should be _____ on a regular basis.
    a. replaced
    b. inspected
    c. lubricated
    d. reeved

29. Placing a rope in sheaves is known as _____.
    a. sheaving
    b. reeving
    c. terminating
    d. winding

30. The easiest method for determining the parts of a line needed for a lift is to _____.
    a. divide the total weight by the nominal breaking strength
    b. check the OSHA regulations
    c. check the load chart
    d. calculate the D/d ratio

31. What is the minimum safety factor on a boom pendant or standing rope for boom erection?
    a. 2.5
    b. 3.0
    c. 3.5
    d. 5.0

32. The safety factor for live or running rope for boom erection is _____.
    a. 2.0
    b. 2.5
    c. 3.0
    d. 3.5

33. All of the following can be used to minimize block twisting *except* _____.
    a. reducing the load
    b. using a longer wire rope
    c. using an even number of parts of a line
    d. using a tag line

34. If you are directed to overwind a right lay rope onto a drum, you would secure the rope on the right side of the drum and move right to left.
    a. True
    b. False

35. The component that actually transfers the power of the crane to the wire rope is the _____.
    a. sheaves
    b. load hook
    c. load block
    d. hoist drum

GLOSSARY

Trade Terms Introduced in This Module

Abrasion: Frictional wear on a wire rope.

Annealed: Heated to reduce stress and increase strength.

Birdcaging: A term used to describe a condition of a wire rope where it is forced into compression and the outer strands are forced away from the core, forming a cage.

Core: The axial part of a wire rope on which the strands are laid.

Corrosion: The chemical breakdown of the materials in a wire rope through an adverse reaction to acids, bases, or moisture.

Dead end: The excess wire rope exposed after some type of end termination has been properly fitted. The part of the wire rope that is not exposed to the strain applied by any type of load, such as a load block, ball, or rigging gear.

Design factor: The ratio of the nominal strength to the total working load for a wire rope.

Drum: A cylindrical barrel on which a wire rope is wound for either storage or operation.

Efficiency: The ratio of a wire rope's actual breaking strength to the aggregate strength of all individual wires tested separately; this is usually expressed as a percentage.

Fatigue: The process of progressive fracture resulting from the bending of individual wires. These fractures may occur at bending stresses well below the ultimate strength of the material. Lack of lubrication and rust on a wire rope can accelerate this process.

Fiber core (FC): A cord or rope of synthetic fiber used as the center of a wire rope.

Filler wire: Small wires within a strand used as spacers that help hold the position and support other wires.

Grade: A wire rope or strand classification by any number of factors, such as type of material and strength.

Headache ball: Ball used in conjunction with the hoisting hook.

Independent wire rope core (IWRC): Wire rope used as the core of a larger rope.

Inner wires: All strand wires except the outer or cover wires.

Kink: A deformation of a wire rope caused by a loop of rope being pulled tight. This causes irreparable damage to the rope.

Lay (length): The distance measured parallel to the axis of the rope or strand in which a strand or wire makes one revolution around the core or center.

Lay (strand): The manner that the wires in a strand of a wire rope are laid in the form of a spiral.

Live section: The part of the wire rope that is exposed to all weight applied to the wire rope.

Load block: A wire rope sheave enclosed in side plates with a lifting attachment (hook or shackle).

Loop: A 360-degree change of direction of a wire rope that results in a kink when pulled tight.

Monel™: A trademark alloy made of copper and nickel.

Nominal breaking strength: The ultimate load at which a wire rope will fail.

Plow steel (PS): A grade of steel used in the manufacture of wire rope.

Preformed: Wires that are permanently formed during fabrication into the helical shape they will be in when placed in a strand.

Reeve: A rope system in which the rope travels around drums and sheaves to change direction or increase mechanical advantage.

Seizing: The securing of an open end of a wire rope by wrapping a wire or strand tightly around its circumference several times. This action prevents fraying and unraveling of the wire rope. This preparation is also used on wire rope prior to cutting it for length from a spool/reel.

Sheave: One of a number of grooved pulleys that guide the wire rope. The types of sheaves may include, but are not limited to, the following types: equalizing, boom tip, idler, and rollers.

Spelter: Molten metal, specifically zinc.

Strand: A grouping of round or shaped wires helically laid around a core.

Strand pattern: Any one of a number of configurations of strand wires used to form a strand. Types of patterns may include, but are not limited to, the following: Seale, Warrington, and filler.

Stress: The load placed on a wire rope divided by the cross-sectional area of the wire.

Stretch: The elongation of a wire rope under load.

Wire: A single, continuous length of metal, cold-drawn or cold-rolled from a rod, available in various cross-sections.

Wire strand core (WSC): A wire strand used as the axial center member of a wire rope.

Figure Credits

Wire Rope Technical Board	201F01 through 201F10, 201F13, 201F14, 201F17, 201F18, 201E01
The Crosby Group, Inc.	201F11, 201F12
Wire Rope Corporation of America, Inc.	201F19

CONTREN® LEARNING SERIES — USER UPDATES

The NCCER makes every effort to keep these textbooks up-to-date and free of technical errors. We appreciate your help in this process. If you have an idea for improving this textbook, or if you find an error, a typographical mistake, or an inaccuracy in NCCER's Contren® textbooks, please write us, using this form or a photocopy. Be sure to include the exact module number, page number, a detailed description, and the correction, if applicable. Your input will be brought to the attention of the Technical Review Committee. Thank you for your assistance.

Instructors – If you found that additional materials were necessary in order to teach this module effectively, please let us know so that we may include them in the Equipment/Materials list in the Instructor's Guide.

Write: Curriculum Revision and Development Department
National Center for Construction Education and Research
P.O. Box 141104, Gainesville, FL 32614-1104

Fax: 352-334-0932

E-mail: curriculum@nccer.org

Craft ______ Module Name ______

Copyright Date ______ Module Number ______ Page Number(s) ______

Description ______

______

______

______

(Optional) Correction ______

______

______

(Optional) Your Name and Address ______

______

______

Module 38202-06

Telescopic Boom Assembly and Disassembly

COURSE MAP

This course map shows all of the modules in the second part of the *Rigging* curriculum, *Rigging*. The suggested training order begins at the bottom and proceeds up. Skill levels increase as you advance on the course map. The local Training Program Sponsor may adjust the training order.

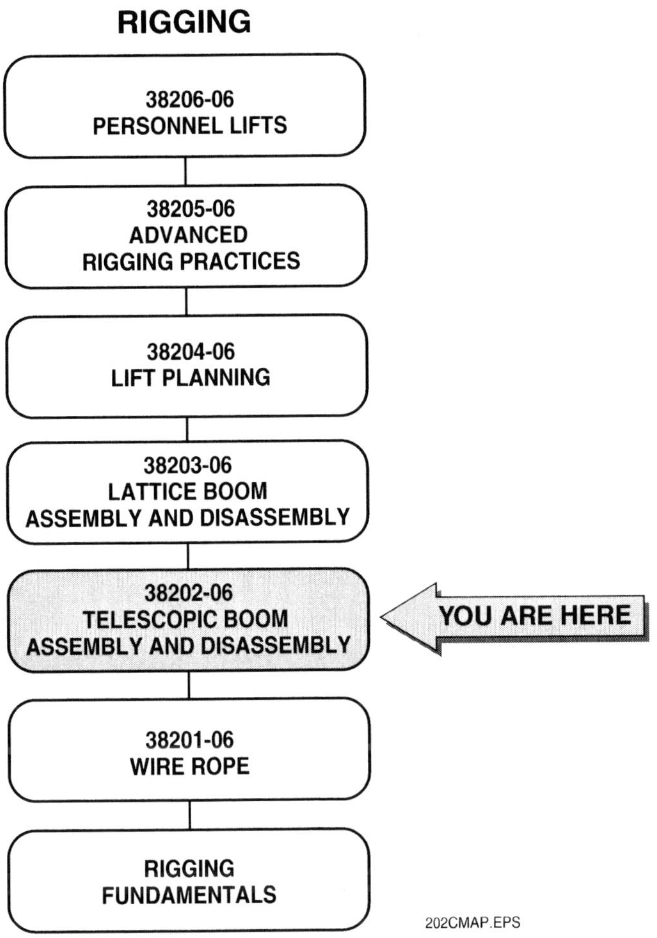

MODULE 38202-06 CONTENTS

1.0.0 INTRODUCTION .2.1

2.0.0 PRE- AND POST-ASSEMBLY CONSIDERATIONS .2.1

3.0.0 COUNTERWEIGHT CONSIDERATIONS .2.1

4.0.0 SWING-AWAY LATTICE EXTENSION .2.2

4.1.0 Installing a Swing-Away Lattice Extension .2.2

4.2.0 Stowing a Swing-Away Lattice Extension .2.6

5.0.0 A-FRAME JIB .2.6

5.1.0 Installing an A-Frame Jib .2.7

5.2.0 Stowing an A-Frame Jib .2.10

6.0.0 INTERMEDIATE BOOM SECTIONS .2.11

6.1.0 Installing Intermediate Boom Sections .2.12

6.2.0 Removing Intermediate Boom Sections .2.12

7.0.0 POWER-PINNED FLY .2.13

7.1.0 Extending a Power-Pinned Fly .2.13

7.2.0 Retracting a Power-Pinned Fly .2.15

8.0.0 AUXILIARY SINGLE-SHEAVE BOOM HEAD .2.15

9.0.0 ROPE INSTALLATION .2.16

SUMMARY .2.16

REVIEW QUESTIONS .2.17

GLOSSARY .2.21

ACKNOWLEDGMENTS .2.23

Figures

Figure 1 Stowed two-piece adjustable fly 2.2
Figure 2 Two-piece adjustable fly assembled at an offset 2.2
Figure 3 Non-adjustable swing-away extension 2.2
Figure 4 Stop rail viewed from under stowed extension 2.3
Figure 5 Stop rail extended under stowed extension 2.3
Figure 6 Swing-away assembly, part one 2.4
Figure 7 Swing-away assembly, part two 2.5
Figure 8 A-frame jib assembly, part one 2.7
Figure 9 A-frame jib assembly, part two 2.8
Figure 10 A-frame jib assembly, part three 2.9
Figure 11 Intermediate boom extensions 2.11
Figure 12 Power-pinned fly extension – manual load pin 2.13
Figure 13 Power-pinned fly extension – pin and gear reducer 2.14
Figure 14 Section two extended with section three inside 2.14
Figure 15 Section three extended with section two retracted 2.14
Figure 16 Auxiliary boom head 2.15
Figure 17 Storage reel to drum 2.16

MODULE 38202-06

Telescopic Boom Assembly and Disassembly

Objectives

When you have completed this module, you will be able to do the following:

1. Determine if there is adequate space and resources for crane assembly and disassembly.
2. Define the relationship of the counterweight to the assembly and disassembly of the boom.
3. Assemble and disassemble various crane attachments and components.

Prerequisites

Before you begin this module, it is recommended that you successfully complete *Rigging Fundamentals* and *Rigging*, Module 38201-06.

Required Trainee Materials

1. Pencil and paper
2. Appropriate personal protective equipment

1.0.0 ◆ INTRODUCTION

Over half of all mobile crane accidents result from human errors made during crane assembly. For this reason, there is a need for the rigger to assist the operator and the site engineer or management in making sure the boom and its attachments are assembled and disassembled properly. Specific guidelines for installing various boom attachments are contained in the operator's safety manual for each crane. The purpose of this module is to give the trainee an overview of the assembly and disassembly procedures for telescoping boom crane attachments.

2.0.0 ◆ PRE- AND POST-ASSEMBLY CONSIDERATIONS

An erection area should be roped off and large enough to allow crane components to be stacked and handled. It should also allow a crane to be assembled without interfering with or posing a hazard to other site personnel. In addition, as the crane swings, there needs to be at least two feet of clearance between the counterweight and the nearest obstacle.

Before assembling the crane, the operator and the person responsible for the lift must consider the following:

- Is there enough room for the crane to maneuver into position on the site?
- Is there a large enough area in which to erect or extend the boom?
- Can the trucks hauling boom sections get into place and be unloaded safely?
- Is there enough timber blocking to support the boom when it is being assembled and disassembled?

3.0.0 ◆ COUNTERWEIGHT CONSIDERATIONS

It is imperative that any personnel involved in the assembly and disassembly of hydraulic boom cranes be aware of counterweight requirements. When attaching boom extensions, the telescope status and superstructure position on some cranes depend on the mounted counterweight. The crane may overturn if the specified maximum telescope status is exceeded or if the position of the superstructure is not observed.

> **WARNING!**
> Assembly and disassembly personnel must always wear the appropriate personal protective equipment and observe safe working practices when working with boom extensions.

4.0.0 ◆ SWING-AWAY LATTICE EXTENSION

A swing-away lattice extension, sometimes called a fly, is designed for the crane with which it is delivered. These extension parts are serial-number specific. The operator should only operate the crane with those parts of the extension which have the same serial number or other identifier as the crane. If the crane is equipped with a load moment indicator (LMI), the indicator is only set for this particular extension. If any other extension is used, the crane will not function properly and damage may occur. *Figures 1* and *2* show a two-piece adjustable fly. This type of swing-away extension can be assembled straight on the boom or at various offsets to the boom. The crane can be used with the extension stowed on the side of the boom if the extension is not required.

4.1.0 Installing a Swing-Away Lattice Extension

The length of non-adjustable swing-away extensions can vary depending on the type of crane and the manufacturer. *Figure 3* shows a rough-terrain crane with only a swing-away extension. Some cranes are equipped with both a swing-away and an A-frame jib on the side of their main booms. A jib attachment adds steps to the swing-away attachment procedure.

202F01.EPS

Figure 1 ◆ Stowed two-piece adjustable fly.

Always follow the manufacturer's instructions and procedures for the crane being used. The specific crane may require different processes than those discussed in this module. The following is a general procedure for installing a non-adjustable swing-away extension. Note that this reflects the product of a particular manufacturer.

> **WARNING!**
> Any personnel involved in boom assembly and installation must observe the warning and safety information provided in the crane manual or site safety manual.

202F02.EPS

Figure 2 ◆ Two-piece adjustable fly assembled at an offset.

202F03.EPS

Figure 3 ◆ Non-adjustable swing-away extension.

Step 1 Extend and set the outriggers or crawlers according to the manufacturer's recommendations. Install proper counterweights as necessary for the required load and boom length.

Step 2 Fully retract all boom sections, and lower the boom to minimum elevation to permit installation of **pins** and access to the boom nose.

Step 3 Rig either the main hoist or optional **auxiliary** hoist cable for a single part of line. Lay the cable off to the left side of the boom while erecting the swing-away.

Step 4 To deploy the stop rail (see *Figure 4*), remove the retaining clip from the stop-rail pin, and pull out the stop-rail pin. Hold the open end of the stop rail with your hand to prevent the stop rail from swinging open.

Step 5 Rotate the stop rail until it is in a horizontal position, and insert the stop-rail pin into the outer bore hole (*Figure 5*). Secure the stop-rail pin in place with its retaining pin. Check that the section is locked with the stop-rail pin and a pivot pin in the bearing point on the stop rail.

For the following steps, refer to *Figure 6*:

Step 6 Ensure the latch lock pins (1 and 2) and latch plates (3 and 4) that secure the A-frame (5 ref) to the front and rear support brackets (6 and 7) are properly engaged.

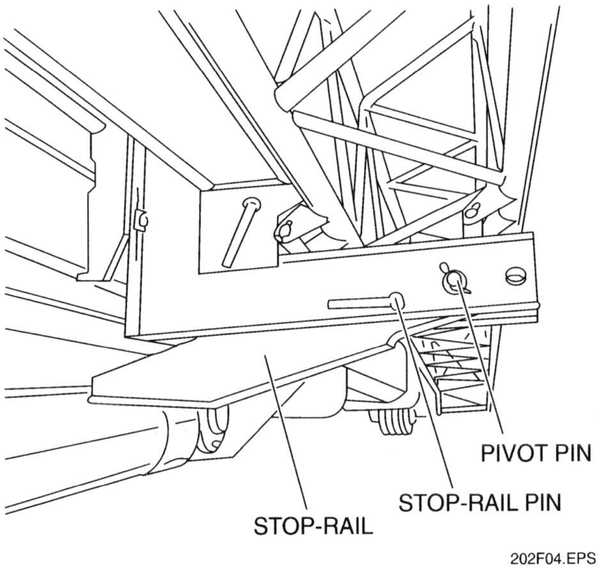

Figure 4 ◆ Stop rail viewed from under stowed extension.

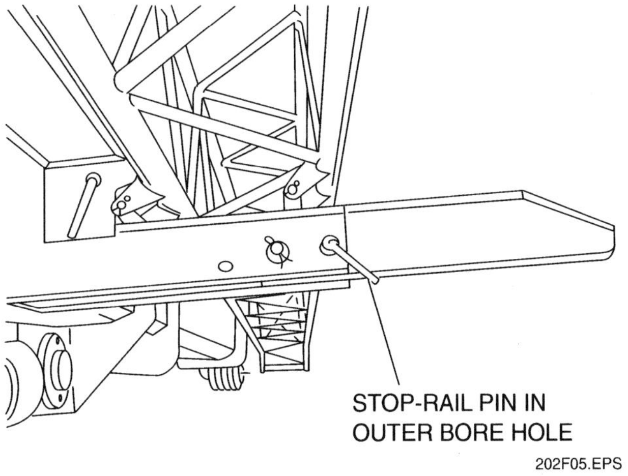

Figure 5 ◆ Stop rail extended under stowed extension.

Step 7 At the base end of the A-frame (5 ref), remove the pin (8) securing the A-frame to the swing-away (9 ref). Stow the pin in the A-frame attaching link (13).

Step 8 At the sheave end of the A-frame (5), remove the pin (10) securing the A-frame to the swing-away. Position the attaching link (11) in the stowed position (12) on the A-frame, and secure it with the pin (10).

Step 9 On the swing-away rear support bracket (7), pull the latch lock pin (15), and swing the latch plate (16) out of the way.

Step 10 With the front support bracket (6) acting as a pivot, pull outward on the swing-away tip to position the swing-away anchor fittings (18) onto the attach fittings (17) on the right side of the boom nose.

Step 11 Remove the attach pins (14) and their retainer clips stowed in the base of the swing-away, and insert them though the attach and anchor fittings on the right side of the boom nose. Install the retainer clips in the pins.

Step 12 Attach a guide rope to the swing-away. This ensures that the swing-away extension swings around safely.

WARNING!

The boom must be raised to horizontal to prevent the swing-away from swinging around after removing the attach pins from the front support bracket. When erecting the swing-away, make sure all personnel and equipment are kept clear of the swing path.

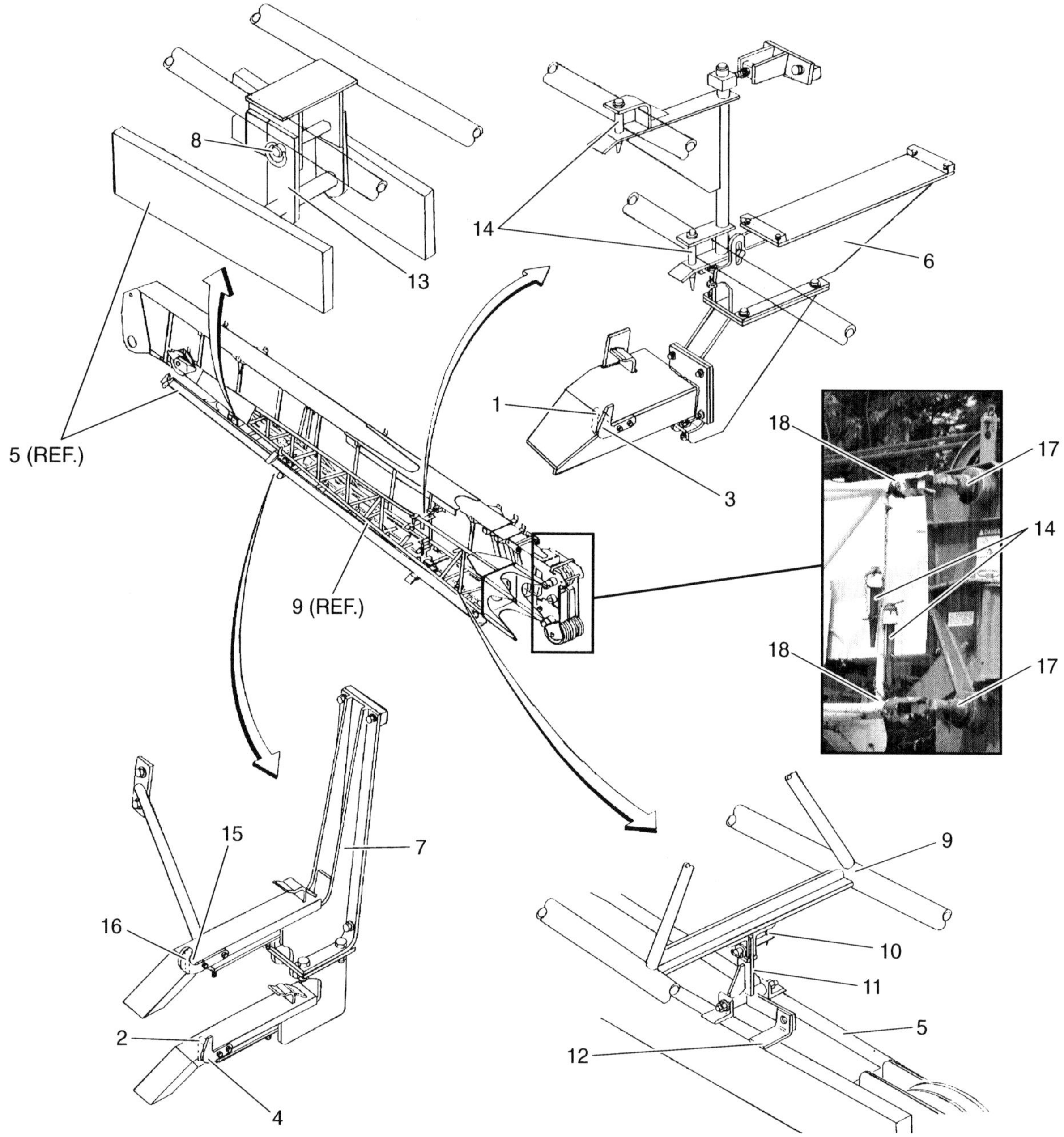

1. LATCH LOCK PIN
2. LATCH LOCK PIN
3. LATCH PLATE
4. LATCH PLATE
5. A-FRAME
6. FRONT SUPPORT BRACKET
7. REAR SUPPORT BRACKET
8. PIN
9. SWING-AWAY
10. PIN
11. ATTACHING LINK
12. ATTACHING LINK STOWAGE POSITION
13. ATTACHING LINK
14. ATTACH PIN
15. LATCH LOCK PIN
16. LATCH PLATE
17. ATTACH FITTINGS
18. ANCHOR FITTINGS

202F06.EPS

Figure 6 ◆ Swing-away assembly, part one.

Step 13 Raise the boom to a level, horizontal position. Then, remove the attach pins (14) and their retainer clips from the front support bracket (6).

For the following steps, refer to *Figure 7*:

Step 14 Swing the swing-away into place in front of the boom nose, engaging the anchor fittings (18) with the attach fittings (17) on the left side of the boom nose.

NOTE

To swing the extension around, use the outriggers to move the crane into a diagonal position and incline the boom slightly. If there is difficulty in aligning the swing-away extension's attach points to the boom nose, assembly personnel should place blocking under the tip of the swing-away to relieve the weight.

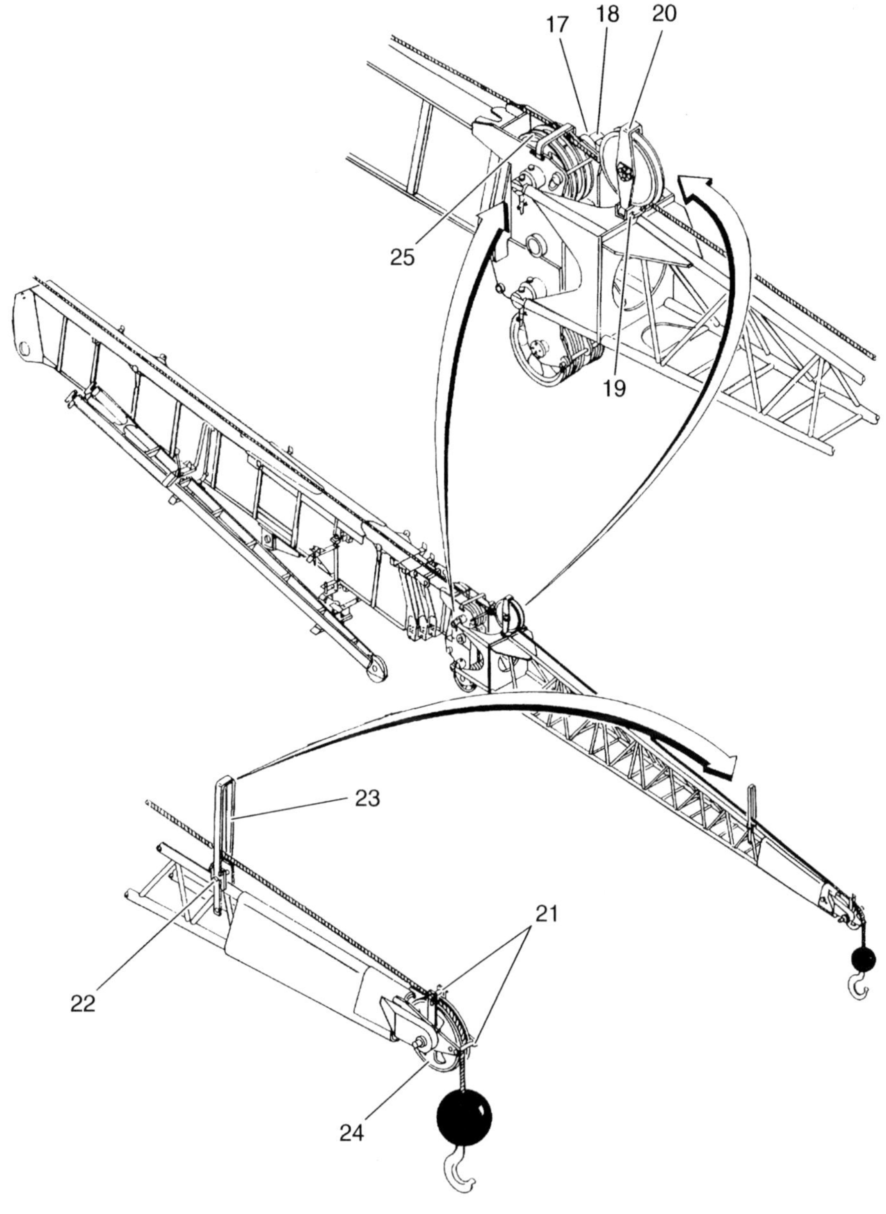

17. ATTACH FITTINGS
18. ANCHOR FITTINGS
19. PIN
20. CABLE-ROLLER SHEAVE
21. CABLE-RETAINING PINS
22. PIN
23. CABLE GUIDE
24. EXTENSION SHEAVE
25. BOOM-HEAD SHEAVE

202F07.EPS

Figure 7 ◆ Swing-away assembly, part two.

Step 15 Lock the swing-away to the main boom by inserting the attach pins (pins removed from the front support bracket) into the anchor (18) and attach fittings (17) on the left side of the boom. Install the retainer clips in the pins.

Step 16 Remove a pin (19) and retainer clip from the cable-roller sheave (20). Move the cable-roller sheave (20) into working position in accordance with the prescribed inclination. Lock it into place with a pin (19), and secure the pin with its retainer clip.

Step 17 Reel off the hoist rope from the main or auxiliary hoist.

Step 18 Lay the hoist cable over the boom-head sheave (25), under the cable-roller sheave (20), and over the extension sheave (24).

Step 19 Remove the pin (22) and retainer clip securing the cable guide (23) in the stowed position. Position the cable guide in the vertical position, with the cable passing through the guide. Reinstall the pin through the second hole in the guide, and install the clip.

Step 20 Make sure that cable-retaining pins (21) are installed.

Step 21 Reeve the hoist cable on the headache ball.

Step 22 Establish electrical connections to the main boom.

Step 23 Insert the anemometer (if equipped) at the head of the swing-away, and establish an electrical connection.

Step 24 Attach the lifting limit switch weight. Feed the hoist rope through the weight.

Step 25 Move the stop rail into transport position.

4.2.0 Stowing a Swing-Away Lattice Extension

Always follow the manufacturer's instructions for stowing a swing-away lattice extension. For general guidelines, you can refer to the following abbreviated procedure:

Step 1 Fully retract the telescopic sections of the boom.

Step 2 Turn the superstructure to the back or the side.

Step 3 Lower the main boom until the load moment indicator (LMI) switches off.

Step 4 Override the LMI.

Step 5 Lower the main boom until the head sheave is approximately five feet above the ground.

Step 6 Unreeve the hoist rope from the load block, and remove it from the head sheave.

Step 7 Remove the lifting limit switch weight.

Step 8 Remove the anemometer (if equipped).

Step 9 Move the main boom into 0-degree position, if necessary.

Step 10 Disconnect the electrical connections to the main boom.

Step 11 Move the cable-roller sheave into the transport position.

Step 12 Wind in the hoist rope.

Step 13 Move the stop rail into assembly position.

Step 14 Attach the guide rope to the swing-away.

Step 15 Remove the connecting pins between the swing-away and main boom on the left-hand side, insert in the holder, and lock.

Step 16 Swing the swing-away on the main boom until the pivot points on the pivot bearing align.

Step 17 Insert pins in the pivot bearing.

Step 18 Remove the connecting pins between the swing-away and main boom on the right-hand side, insert in the holder, and lock.

Step 19 Move the stop rail into transport position.

5.0.0 ◆ A-FRAME JIB

The A-frame jib consists of two side rails and cross members made of rectangular tubing and welded together to form an A-frame. On some models of cranes, it is the only boom extension attachment. On others, it is attached to the swing-away and will be extended with the swing-away. This section provides a procedure for one particular crane. Be sure to follow the procedure recommended by the manufacturer of your crane.

5.1.0 Installing an A-Frame Jib

When erecting or stowing the jib, personnel must follow the manufacturer's instructions and the warning decals installed on the swing-away and/or A-frame. A general procedure to install an A-frame jib is as follows:

Step 1 Extend and set the outriggers or crawlers according to the manufacturer's recommendations.

Step 2 Swing the boom to the proper operational quadrant.

> **WARNING!**
> The boom must be completely retracted before engaging the swing-away boom extension with the boom nose.

Step 3 Fully retract all boom sections, and lower the boom to minimum elevation.

Step 4 Rig either the main hoist or optional auxiliary hoist cable for single part of line with nothing but the **becket** on the end of the cable.

For the following steps, refer to *Figure 8*:

> **WARNING!**
> Do not release the A-frame latch plates unless the pins (2 and 6) attaching the A-frame to the swing-away are in place.

Step 5 Pull the latch lock pins on the front and rear support brackets, and swing the A-frame latch plates out of the way.

Step 6 Assemble the swing-away extension to the boom as described in a previous section, but do not release the A-frame jib from the swing-away extension.

Step 7 Telescope the boom, and elevate it as necessary to prevent damage to the swing-away nose.

Step 8 Reeve the hoist cable over the boom nose center **idler** sheave, down between the boom nose and the base of the swing-away, and connect the becket to the cable-retaining pin (1) on the A-frame nose.

Step 9 Slowly reel in the hoist cable to relieve pressure on the pin (2) securing the A-frame nose to the base of the swing-away.

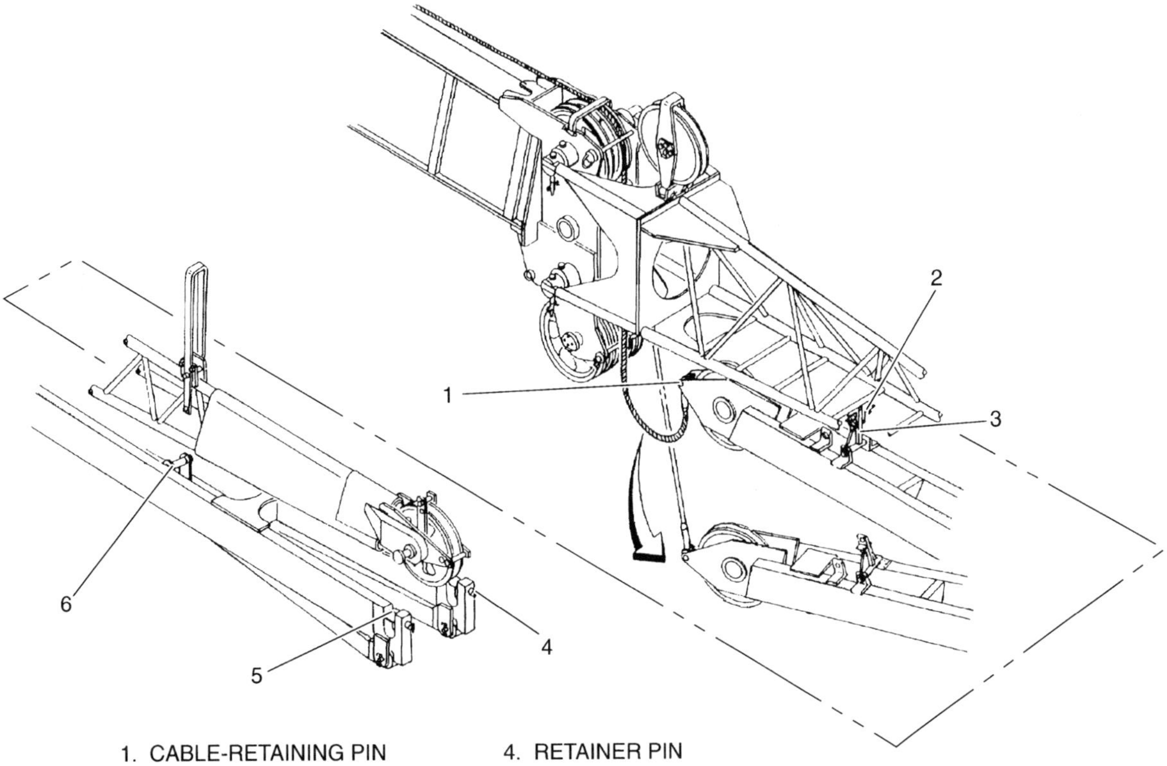

Figure 8 ◆ A-frame jib assembly, part one.

Step 10 Remove the pin (2) and its retainer clip. Position the attaching link (3) in the stowed position on the A-frame, and secure it with the pin. Install the pin's retainer clip.

Step 11 Remove the retainer pins (4), their clips, and the retainer **bushings** (5) from the attach fittings on the base of the A-frame.

Step 12 Slowly reel out the hoist cable, allowing the A-frame to pivot on the pin (6) in the front attach link until the A-frame attach fittings engage with the shaft on the swing-away nose. Install the retainer bushings (5) and retainer pins (4). Install the pin clips.

Step 13 Lift the A-frame jib nose up slightly, and remove the pin (6) and its retainer clip, securing the base of the A-frame to the swing-away. Stow the pin in the swing-away, and secure the pin with its retainer clip.

Step 14 Slowly reel out the hoist cable, and lower the A-frame nose to the ground. Disconnect the hoist cable becket from the cable retainer pin (1) in the A-frame nose.

Step 15 Remove the hoist cable from between the boom nose and the swing-away.

Step 16 Elevate the boom until the A-frame is perpendicular to the ground.

Step 17 With the A-frame touching the ground, retract and lower the boom so that the A-frame extends out from the end of the swing-away. It will be necessary to reel in the hoist cable while lowering and retracting the boom.

For the following steps, refer to *Figure 9*:

Step 18 Install offset links (8), using pins (7), in the correct fittings on each side of the boom nose applicable to different degrees of offset. Secure pins (7) with their retainer clips.

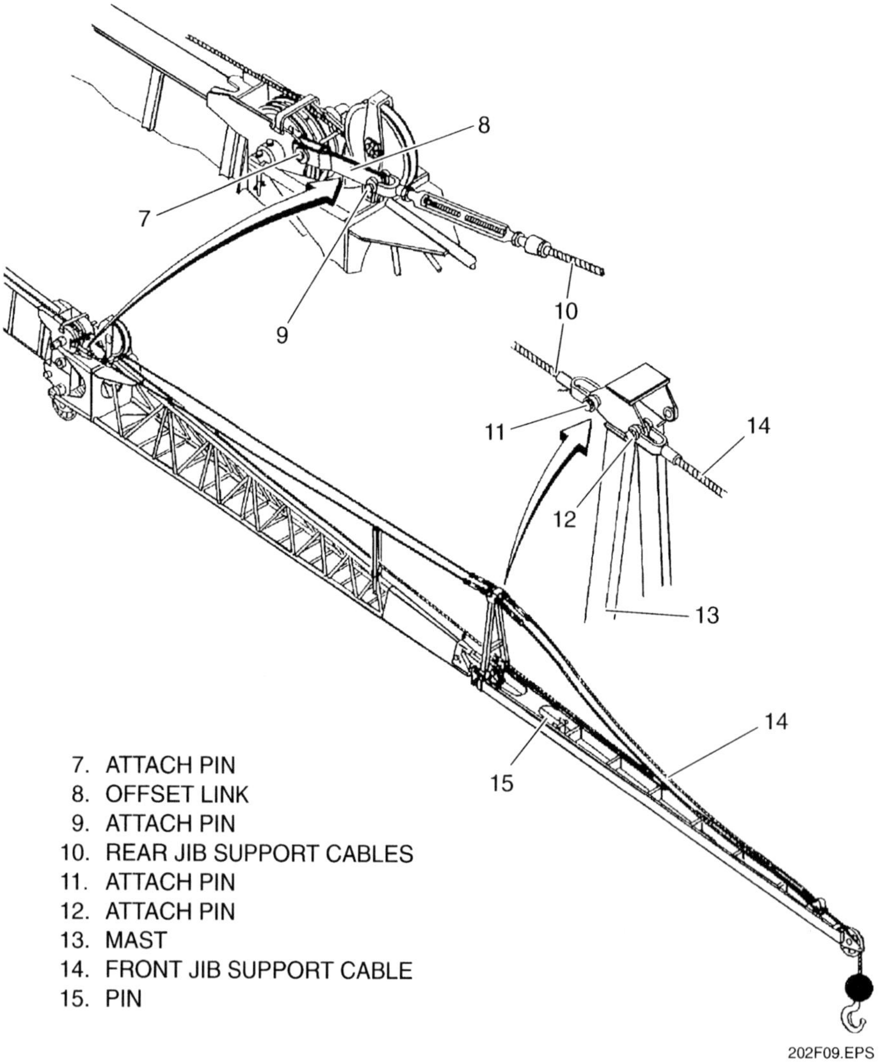

Figure 9 ◆ A-frame jib assembly, part two.

Step 19 Unhook the front jib support cables (14) from their stowed position.

Step 20 Pull the pin (15) and its retainer clip, latching the **mast** (13) to the A-frame.

Step 21 Reel out the hoist cable, and attach the becket to the **lug** on the mast (13).

Step 22 Reel in the hoist cable, raising the mast. Connect the rear jib support cables (10) to the mast using attach pins (11) and to the offset links (8) using attach pins (9). Secure pins with their retaining clips.

Step 23 Reel out the hoist cable, and disconnect the hoist cable becket from the lug on the mast (13).

Step 24 Set the required jib offset by adjusting the rear jib support cable (10) **turnbuckles**.

For the following steps, refer to *Figure 10*:

Step 25 Reeve the hoist cable over the boom nose center idler sheave, over the extension sheave, and over the A-frame sheave (24).

Step 26 Remove the pin (19) and its retainer clip from the stowed position on the swing-away. Raise the cable roller (20) to an upright position so the hoist cable is under the sheave. Reinstall the pin (19) through the second hole in the guide, and secure the pin with its retainer clip.

Step 27 Remove the pin (22) and its retainer clip, securing the cable guide (23) in the stowed position. Position the cable guide in the vertical position, with the cable passing through the guide. Reinstall the pin through the second hole in the guide, and secure the pin with its retainer clip.

Step 28 Make sure the cable-retaining pin (21) is installed.

Step 29 Rig the hoist cable.

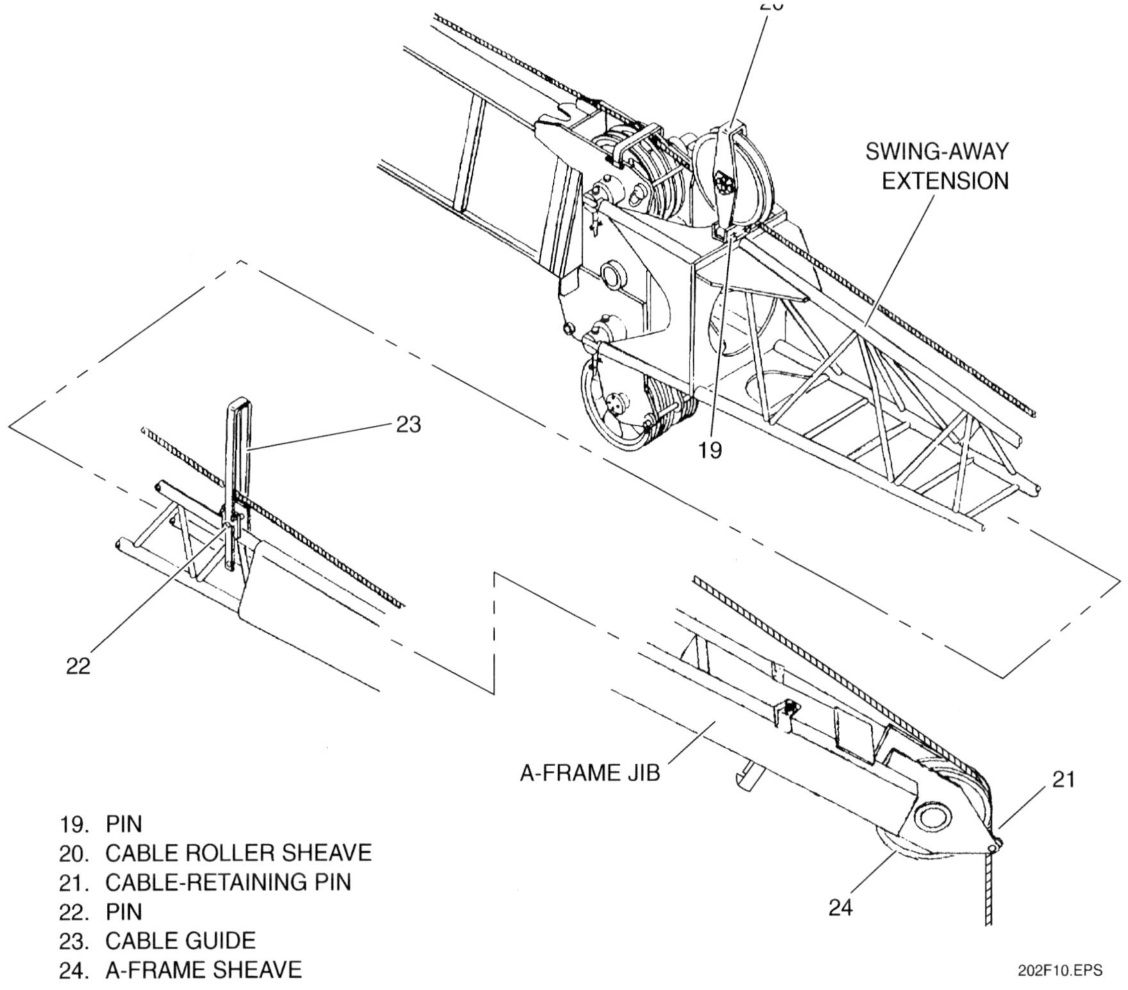

Figure 10 ◆ A-frame jib assembly, part three.

5.2.0 Stowing an A-Frame Jib

Once operations using the jib are completed, it should be stowed. The following are abbreviated steps for stowing an A-frame jib (refer to *Figure 10*), but be sure to follow the manufacturer's procedures for a specific crane:

Step 1 Fully retract the boom. Swing the boom to the appropriate quadrant with the outriggers extended and set and with the crane level.

Step 2 Lower the boom to minimum elevation, and remove the load block or ball.

Step 3 Remove the pin (22) and its retainer clip, securing the cable guide (23). Position the cable guide in the stowed position, and reinsert the pin. Secure the pin with the retainer clip.

Step 4 Remove the pin (19) and its retainer clip, securing the cable roller sheave (20). Lay the cable roller over, and reinsert the pin in the bracket. Secure the pin with its retainer clip.

Step 5 Remove the cable-retaining pin (21) from the A-frame sheave (24). Reel in the hoist cable, removing it from the extension sheave and boom nose idler sheave. Lay the cable off to the side of the boom.

For the following steps, refer to *Figure 9*:

Step 6 Reel out the hoist cable, and connect the hoist cable becket to the lug on the mast (13).

Step 7 Reel in the hoist cable until pressure is relieved from the rear jib support cables (10).

Step 8 Remove the rear jib support cables (10) from the mast (13).

Step 9 Reel out the hoist cable until the mast (13) is laying on the A-frame.

Step 10 Remove the hoist cable becket from the mast lug, and reel in the cable.

Step 11 Secure the mast to the A-frame with the pin (15). Install the pin retainer clip.

Step 12 Stow the front jib support cable (14) in the stowage brackets.

Step 13 Remove the rear jib support cables (10). Also remove the offset links (8) from the fittings on the boom nose if they have been used.

Step 14 Reeve the hoist cable over the boom nose center idler sheave and down between the boom nose and base of the swing-away.

Step 15 Elevate and extend the boom until the A-frame is perpendicular to the ground and the A-frame nose is touching the ground.

For the following steps, refer to *Figure 8*:

Step 16 Reinstall the cable-retaining pin (1) in the A-frame nose. Reel out the hoist cable, and connect the hoist cable becket to the cable retainer pin (1).

Step 17 Reel in the hoist cable, and lower and retract the boom until the A-frame is pulled up under the swing-away.

Step 18 Remove the attach pin (6) and its retainer clip from the base of the A-frame, and attach the base of the A-frame to the nose of the swing-away using the pin. Secure the pin with its retainer clip. Then, unreel the hoist cable slightly.

Step 19 Remove the retainer pins (4), their retainer clips, and the retainer bushings (5) from the fittings on the base of the A-frame.

CAUTION

Make sure the pins and retainer bushings are removed from the base of the A-frame before reeling in the hoist cable. Damage to the A-frame could occur if they are not.

Step 20 Reel in the hoist cable until the attaching link (3) on the A-frame nose can be attached to the swing-away. Remove the pin (2) and its retainer clip, and swing the attaching link (3) up. Install the pin, and secure it with its retainer clip.

Step 21 Reinstall the retainer bushings (5) and retainer pins (4) in the A-frame base fittings. Install the pin retainer clips.

Step 22 Remove the hoist cable becket from the A-frame nose, and reinstall the cable retainer pin (1).

Step 23 Fully retract the boom.

Step 24 Stow the swing-away.

Step 25 Swing the A-frame latch plates on the front and rear support brackets up, and secure them with the latch lock pins. This secures the A-frame to the brackets.

Step 26 Rig the boom nose and hoist cable as desired, and operate the crane using normal operating procedures.

6.0.0 ◆ INTERMEDIATE BOOM SECTIONS

The boom extension of a hydraulic crane may also include **intermediate** boom sections (see *Figure 11*). This portion of the module will cover intermediate boom sections and the procedures for installing them on one particular model crane.

Before installing the boom extensions, you must establish the rigging mode of the crane. As with the swing-away and A-frame jib, the outriggers must be extended and set, the appropriate amount of counterweight used, and the main boom completely retracted and lowered into the horizontal position.

WARNING!

The boom extension may only be mounted with the span and counterweight indicated in the lifting capacity table. Otherwise, the assembly personnel risk overturning the crane.

Assembly and disassembly personnel must follow the manufacturer's instructions and only use authorized rigging with sufficient loadbearing capacity when assembling boom extensions.

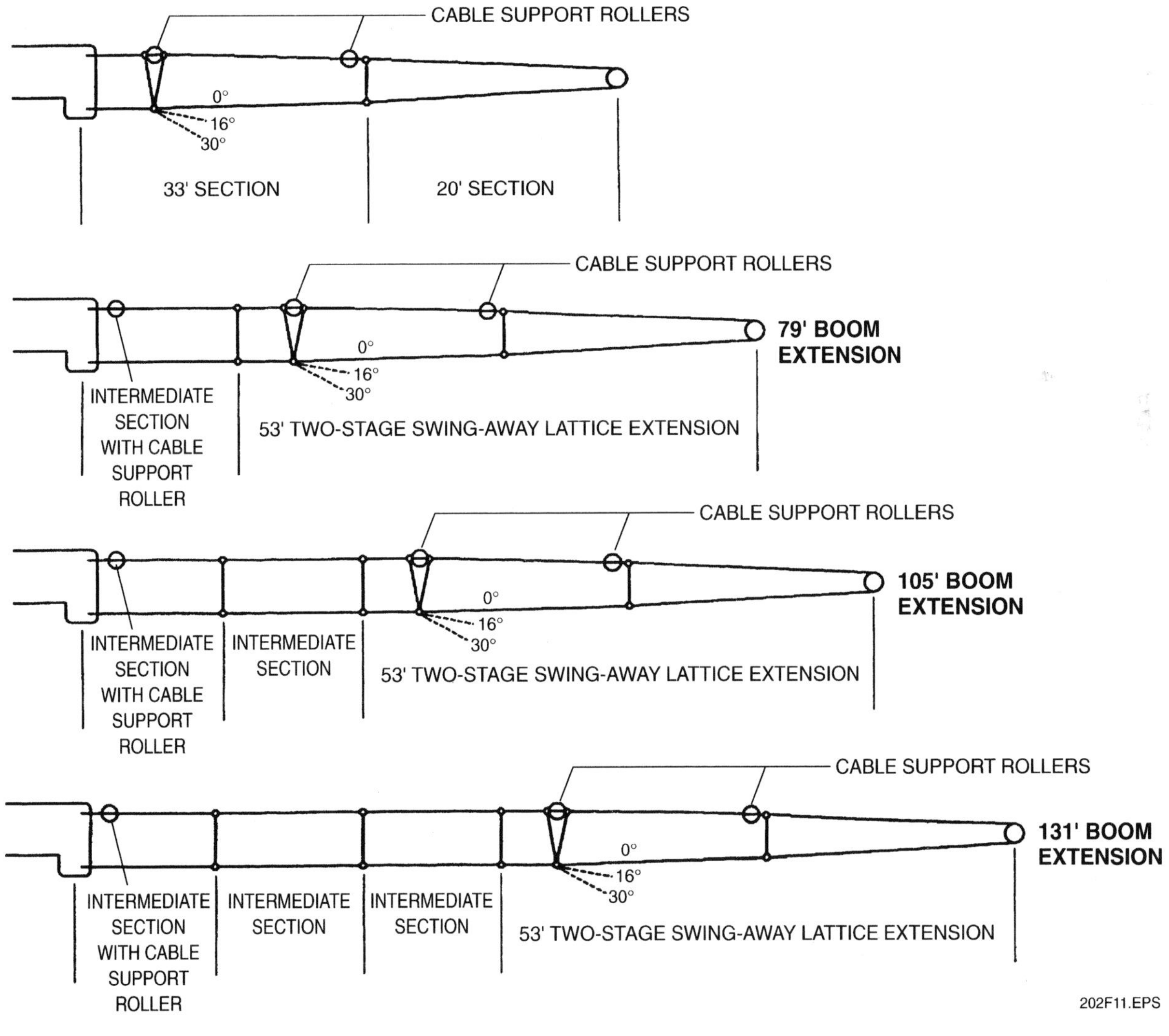

Figure 11 ◆ Intermediate boom extensions.

6.1.0 Installing Intermediate Boom Sections

Use the following steps and an auxiliary crane when attaching intermediate boom sections to the main boom. This abbreviated procedure is for a particular manufacturer's crane. Always consult the manufacturer's guidelines for the crane being used.

Step 1 Before mounting the intermediate boom sections, detach the swing-away and A-frame jib from the main boom, and place them on the ground using an auxiliary crane.

Step 2 Attach the rigging on the intermediate section with the cable support roller. Attachment devices are provided on the intermediate sections for this purpose.

Step 3 Use the auxiliary crane to raise the intermediate section with the cable support roller in front of the boom head until the bores on the intermediate section align with the bearing points on the boom head.

CAUTION

Always assemble the intermediate section with the cable support roller directly in front of the main boom head. The hoist rope may be damaged on sections of the main boom head if the intermediate section is assembled without the cable support roller.

Step 4 Remove the pins from the holders on the intermediate section, and insert them at the top and bottom in the bearing points between the boom head and the intermediate section. Fasten the pins using retaining clips.

Step 5 Use an auxiliary crane to raise the swing-away in front of the intermediate section until the bores on the foot section of the swing-away align with the bearing points on the intermediate section.

Step 6 Remove the pins from the holders on the swing-away, and insert them at the top and bottom in the bearing points between the intermediate section and the swing-away. Fasten the pins using retaining clips.

Step 7 If an offset is required, use the auxiliary crane to support the end of the assembled extension. Then remove the offset adjustment pins, and lower the end of the extension until the pins can be reinserted for the proper offset. Secure the pins with their retaining clips.

Step 8 Complete all electrical connections.

Step 9 Place the hoist cable through the anti-two-blocking switch weight, and reeve the hook ball or load block.

6.2.0 Removing Intermediate Boom Sections

When operations are completed and the intermediate boom sections are no longer needed, follow the abbreviated procedure in this section to remove the extensions. An auxiliary crane must be used to disassemble and incline the intermediate boom sections.

- Retract all telescopic sections completely with the boom raised.
- Turn the superstructure to the appropriate quadrant, depending on the space available.
- Lower the main boom until the LMI triggers a shutdown.
- Override the LMI, and lower the main boom into the horizontal position.

If the boom extension is offset, proceed as follows:

Step 1 Lower the main boom until the head sheave on the offset boom extension is just resting on a board on the ground.

Step 2 Remove the pins for offset adjustment from the bearing points.

Step 3 Raise the end of the boom extension using the auxiliary crane.

Step 4 Move the boom extension into the 0-degree offset position with the auxiliary crane, and pin the offset adjusting system in this position. Then perform the following steps for a non-offset boom extension.

If the boom extension is not offset, proceed as follows:

Step 1 Remove the anti-two-blocking switch weight from the hoist cable on the boom extension, and unreeve the ball or load-block hook.

Step 2 Remove the hoist cable from all sheaves on the boom extension.

Step 3 Remove the retaining pin locking the anti-two-block switch from the swing-away.

Step 4 Release the anti-two-block switch lock on the main boom.

Step 5 Disconnect the electrical connection at all connecting points between the distribution boxes at the main boom and the swing-away.

Step 6 Spool the cable onto the appropriate devices.

Step 7 Screw the plugs of the connecting cable onto the dummy sockets of the individual parts of the boom extension.

Step 8 Screw the short-circuit plugs onto the distribution boxes on the main boom and on the swing-away.

Step 9 Screw the dummy plugs onto the dummy socket on the intermediate boom sections.

Step 10 Attach the appropriate lifting gear to the swing-away, and hold it with the auxiliary crane.

Step 11 Remove the pins between the swing-away and the intermediate boom section, and insert the pins in the bracket on the swing-away. Secure the pins with their retaining clips.

Step 12 Place the swing-away on the ground.

Step 13 Attach the lifting tackle to the intermediate boom section, and hold the section with the auxiliary crane.

Step 14 Remove the pins securing the intermediate boom section, and insert the pins into the bracket on the intermediate boom section. Secure the pins with their retaining clips.

Step 15 Place the intermediate boom section on a separate vehicle.

7.0.0 ◆ POWER-PINNED FLY

The power-pinned fly is another type of boom extension. It is used on hydraulic telescoping-boom cranes. The following sections cover steps for extending and retracting a particular manufacturer's power-pinned fly.

7.1.0 Extending a Power-Pinned Fly

This is a procedure for a particular manufacturer's crane. Always follow the manufacturer's instructions for the crane being used. Follow these steps for extending a power-pinned fly:

Step 1 Position the machine to allow the boom to be extended with the boom over the front, side, or rear of the carrier.

Step 2 Extend the outriggers, and level the machine.

Step 3 Fully retract the boom, and make sure that the manual load pin is engaged in the appropriate section (*Figure 12*).

Step 4 Move the manual extension switch to the ON position. The switch is located in the upper cab. This will energize the solenoid on the flow divider, which blocks flow to the lower telescope cylinder and locks the lower section.

Step 5 Push the toe of the boom extend pedal until the upper telescope cylinder is fully extended. The rear clearance hole in section two and the manual extension pin hole in section three should now be aligned with the access holes in the base section and section one. The pins on the end of the manual extension gear reducer must now be visible when viewed through the access holes in the boom sections (*Figure 13*).

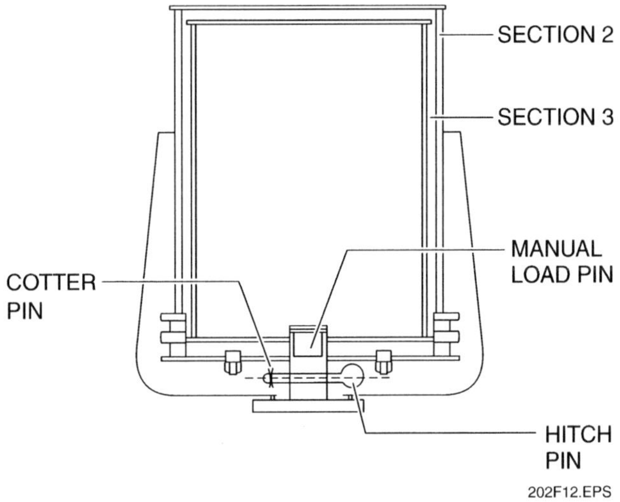

Figure 12 ◆ Power-pinned fly extension – manual load pin.

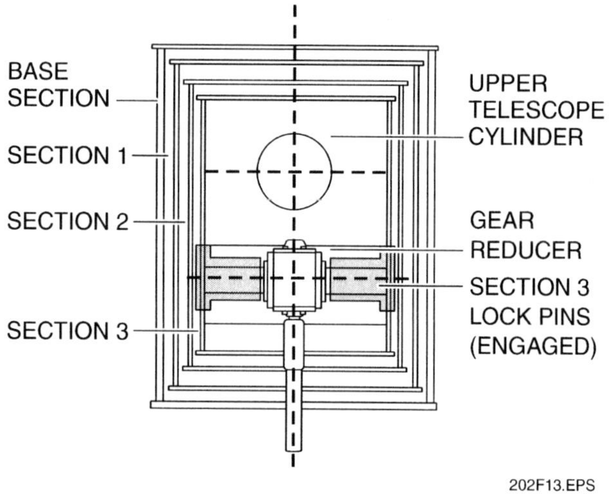

Figure 13 ◆ Power-pinned fly extension – pin and gear reducer.

Step 6 Insert the hand crank through the opening in the bottom of the boom, and turn the input shaft of the gear reducer clockwise to extend the gear reducer lock pins into section three (*Figure 14*). This will pin section three to the lower telescope cylinder. Visually inspect through the access holes to ensure that the lock pins are fully extended into section three. Remove the crank when the pins are fully extended.

Step 7 Disengage the manual load pin by pushing up on the bottom of the pin and removing the cotter pin and the hitch pin (refer to *Figure 12*).

CAUTION

Do not retract the upper telescope cylinder until the manual load pin is fully disengaged from section three. Boom components could be damaged if the load pin is not fully disengaged.

Step 8 Fully retract the upper telescope cylinder by depressing the heel of the boom telescope pedal. Section two will fully retract, and section three will stay extended since it is pinned to the end of the lower telescope cylinder. When fully retracted, the manual load pin in section two will align with the hole located at the base of section three.

Step 9 Engage the manual load pin by pushing up on the bottom of the pin and inserting the hitch pin and cotter pin (refer to *Figure 12*).

Step 10 Insert the hand crank through the opening in the bottom of the boom, and turn the input shaft of the gear reducer counterclockwise to retract the lock pins from section three (*Figure 15*). This will unpin section three from the lower telescope cylinder. Make a visual inspection through the access holes to ensure that the lock pins are fully retracted from section three. Remove the crank when the pins are fully retracted.

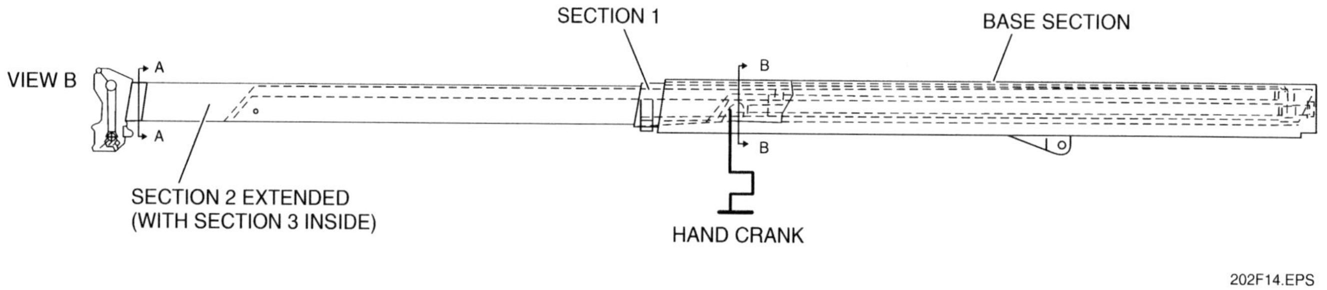

Figure 14 ◆ Section two extended with section three inside.

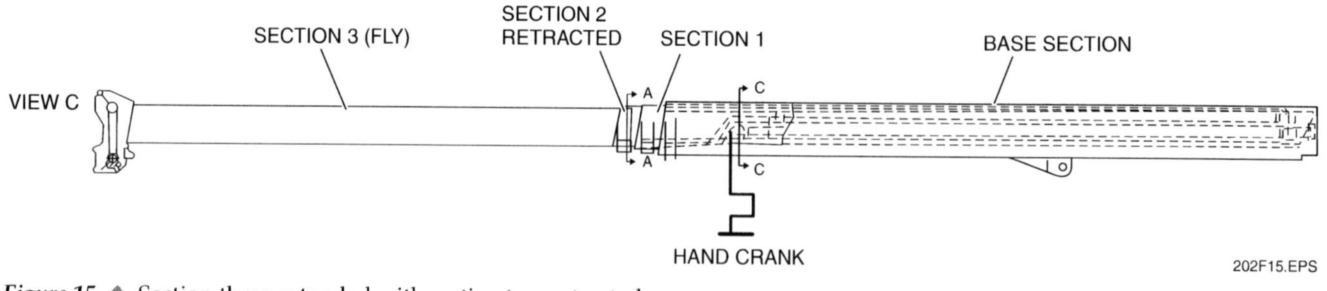

Figure 15 ◆ Section three extended with section two retracted.

Step 11 Move the manual extension switch to the OFF position. This will disengage the solenoid on the flow divider to return equal flow to the telescope cylinders for synchronous operation.

Step 12 The manual section three is now extended and pinned to section two. Powered sections one and two will operate normally.

7.2.0 Retracting a Power-Pinned Fly

Once operations involving the power-pinned fly are completed, it should be retracted and secured. To retract the power-pinned fly, follow these steps:

Step 1 Extend the outriggers, and level the machine.

Step 2 Fully retract the boom to align the pin holes in section three with the lock pins on the gear reducer.

Step 3 Insert the hand crank through the opening in the bottom of the boom, and turn the input shaft of the gear reducer clockwise to extend the gear reducer lock pins into section three (refer to *Figure 15*). This will pin section three to the lower telescope cylinder. Visually inspect through the access holes to ensure that the lock pins are fully extended into section three. Remove the crank when the pins are fully extended.

Step 4 Disengage the manual load pin by pushing up on the bottom of the pin (refer to *Figure 12*). Remove the cotter pin and hitch pin.

Step 5 Move the manual boom extension switch to the ON position. The switch is located in the upper cab. This will energize the solenoid on the flow divider, which blocks flow to the lower telescope cylinder and locks the lower section.

Step 6 Depress the toe of the boom telescope pedal until the upper telescope cylinder is fully extended.

Step 7 Engage the manual load pin by pushing up on the bottom of the pin and inserting the hitch pin and cotter pin (refer to *Figure 12*).

Step 8 Insert the hand crank through the opening in the bottom of the boom, and turn the input shaft of the gear reducer counterclockwise to retract the lock pins from section three (refer to *Figure 14*). This will unpin section three from the lower telescope cylinder. Make a visual inspection through the access holes to ensure that the lock pins are fully retracted from section three. Remove the crank when the pins are fully retracted.

CAUTION
Do not retract the upper telescope cylinder until the lock pins are fully disengaged from section three. Boom components could be damaged if the lock pins are not fully disengaged.

Step 9 Fully retract the upper telescope cylinder by depressing the heel of the boom telescope pedal. The boom is now in the fully retracted position.

Step 10 Move the manual extension switch to the OFF position. This will disengage the solenoid on the flow divider to return equal flow to the telescope cylinders for synchronous operation.

Step 11 The manual section is retracted and pinned to section two. The powered sections operate normally.

8.0.0 ◆ AUXILIARY SINGLE-SHEAVE BOOM HEAD

The auxiliary single-sheave boom head (*Figure 16*) is an additional piece of equipment that can be installed on the boom. In order to attach it to the boom, the crane must be on outriggers or crawlers, the superstructure must be turned to a suitable position, and the main boom must be retracted and set horizontal. To install an auxiliary single-sheave boom head, use the steps that follow.

202F16.EPS

Figure 16 ◆ Auxiliary boom head.

Step 1 Swing the boom top in front of the end of the last boom section. Insert a securing pin through the upper and lower bearing points, and secure it with a retaining clip.

Step 2 Screw the short-circuit plug into the dummy socket on the boom head.

Step 3 Remove the rope guide bars from the head of the main boom and the boom head.

Step 4 Lay the hoist rope over the boom head and over the appropriate sheaves.

Step 5 Reinsert the rope guide bars, and secure them with retaining pins.

Step 6 Hang the hook tackle on the auxiliary hoist rope.

Step 7 Suspend a lifting limit switch weight both on the lifting limit switch on the auxiliary single-sheave boom head and on the lifting unit switch on the main boom.

Step 8 Guide the hoist rope through the lifting limit switch weight.

9.0.0 ◆ ROPE INSTALLATION

The primary concern when installing wire rope is getting the rope on the equipment without trapping any twist that may have been induced during handling or installation. The preferred technique for installing rope onto a crane is as follows:

Step 1 Unload the rope properly, and relieve any twists.

Step 2 Attach the end of the rope to the drum. Pull the rope over the point sheave and attach the end to the drum.

Step 3 Spool the rope onto the underside of the drum slowly and carefully (*Figure 17*).

Step 4 Spool the first layer tightly. The first layer forms the foundation, or grooving, for the remaining upper layers. If it is not tight, wraps in upper layers will pull down between wraps already on the drum. This can lead to rope damage and reduced rope strength and service life.

Step 5 Spool multiple layers with sufficient tension. The tension load should range from one to two percent of the rope's nominal strength.

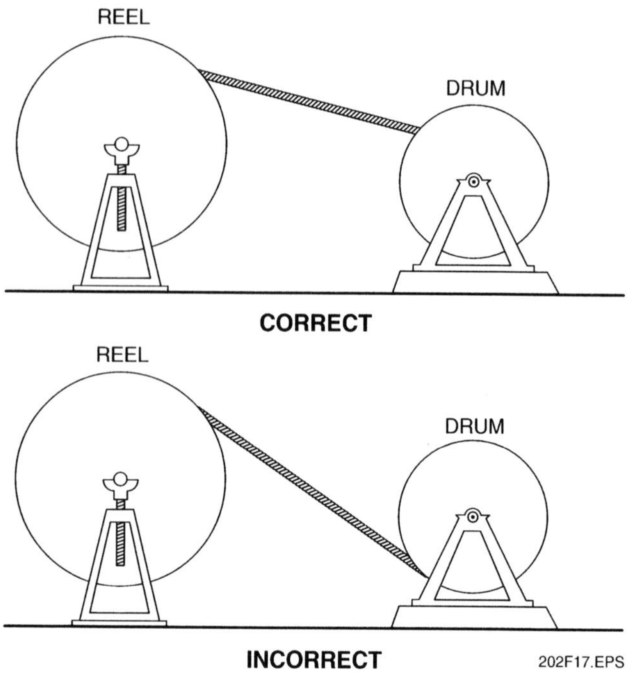

Figure 17 ◆ Storage reel to drum.

Step 6 If the ropes are in a multi-part system, reeve the traveling block and boom tip sheaves so that the rope spacing is maximized. This is done so the traveling block hangs straight and level, and to assure block stability.

Step 7 Check the rope for twists.

Step 8 Cycle the new rope several times, allowing it to form itself to the drum before placing it into service.

Summary

Lifting operations depend on proper assembly of the boom of the crane. On telescopic cranes, this can involve the proper extension of a power-pinned fly or the proper attachment of the swing-away, A-frame jib, and any intermediate boom extensions to the main boom. Assembly and disassembly personnel must be aware of clearance requirements on the job site and must follow the appropriate safety requirements associated with this work.

The procedures given in this module are general steps only; always adhere to the manufacturer's instructions when performing attachment assembly.

Review Questions

1. Work site personnel can find information for installing various boom attachments on a particular crane _____.
   a. on the boom parts
   b. on the crane manufacturer's website
   c. on the load charts
   d. in the operator's safety manual

2. Assembly errors cause _____ of all crane accidents.
   a. half
   b. more than half
   c. less than half

3. Proper counterweights are important in boom assembly because _____.
   a. the crane could overturn during assembly
   b. extra boom attachments, not specific to the crane, can then be attached
   c. they eliminate the need for proper ground support
   d. they allow room for error in assembly

4. When the swing-away extension covered in this module is assembled to the boom, the _____ is used as the first pivot point.
   a. stop rail
   b. attach fitting
   c. front support bracket
   d. A-frame bracket

5. During attachment of the swing-away extension covered in this module, the electrical connections should be established to the main boom immediately after _____.
   a. reeling off the hoist rope from the main hoist
   b. laying the hoist rope over both head sheaves
   c. reeving the hoist rope on the headache ball
   d. attaching the lifting limit switch weight

Using *Figure 1*, match the callout numbers on the figure to the correct descriptions.

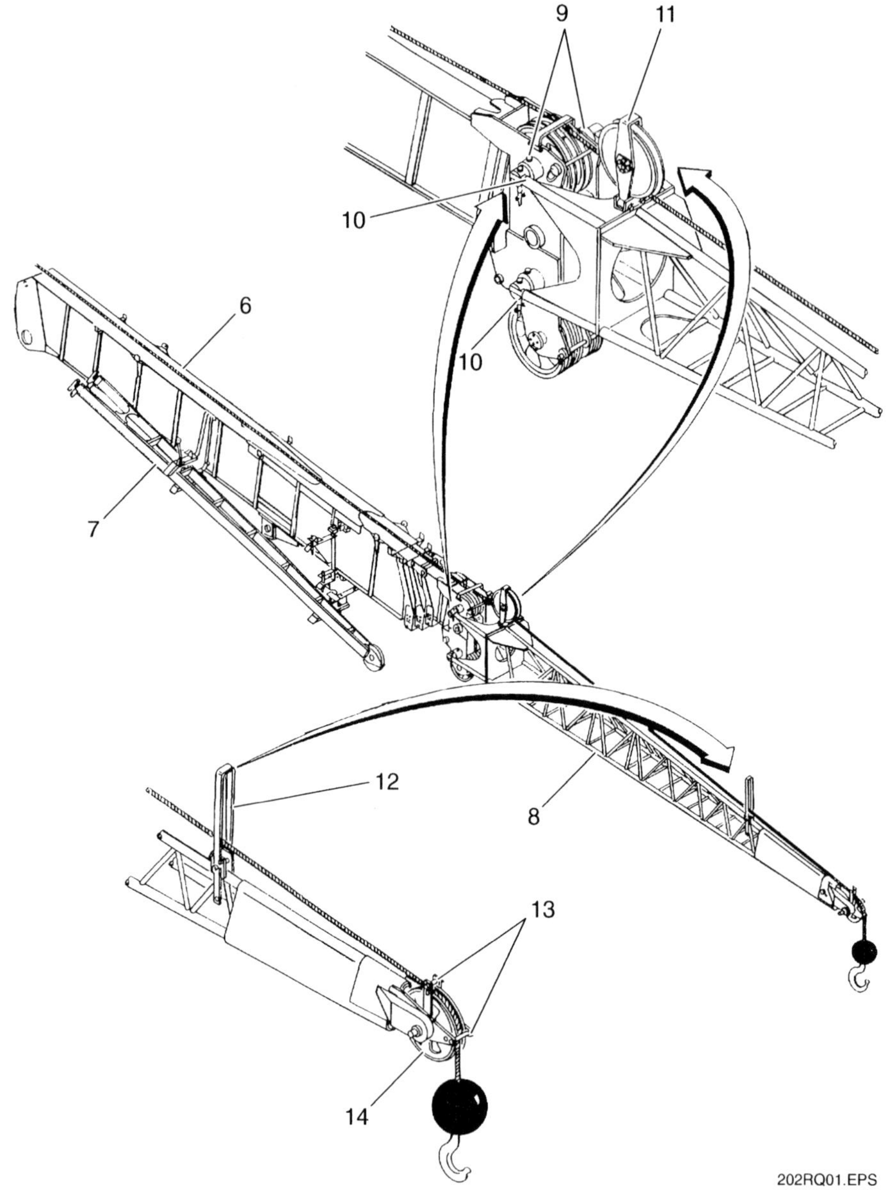

Figure 1

6. _____
7. _____
8. _____
9. _____
10. _____
11. _____
12. _____
13. _____
14. _____

a. Retainer bushing and pin
b. Swing-away extension
c. Cable-retaining pins
d. Attach fitting
e. Cable-roller sheave
f. A-frame jib
g. Telescoping boom
h. Anchor fitting
i. Attaching link
j. Cable guide
k. Extension sheave

15. The first step in stowing the swing-away extension is to _____.
    a. lower the main boom until the LMI switches off
    b. override the LMI
    c. fully retract the telescopic sections of the boom
    d. turn the superstructure to the back or to the side

16. The A-frame jib covered in this module is assembled to the nose of the swing-away extension using _____.
    a. offset links
    b. turnbuckles
    c. retainer bushings and pins
    d. an adjustable jib mast

17. The final offset for the A-frame jib covered in this module is made using _____.
    a. offset links
    b. turnbuckles
    c. retainer bushings and pins
    d. an adjustable jib mast

18. A replaceable steel tube mounted in a case or housing as a bearing is called a(n) _____.
    a. becket
    b. bushing
    c. idler
    d. lug

19. A(n) _____ sheave is used to equalize tension in opposite parts of a rope.
    a. idler
    b. turnbuckle
    c. pin
    d. lug

20. When stowing the A-frame jib covered in this module, tension on the rear jib support cables is relieved by _____.
    a. reeling in a hoist cable attached to the jib mast
    b. releasing the forward jib support cables
    c. loosening the turnbuckles
    d. using supports under the A-frame jib

21. Assembly and disassembly personnel must only use _____ with sufficient load-bearing capacity when assembling boom extensions.
    a. protective equipment
    b. authorized rigging
    c. non-stick lubrication
    d. operator aids

22. For the intermediate boom extensions covered in this manual, any intermediate boom extension can be assembled to the boom head.
    a. True
    b. False

23. When removing intermediate boom sections, if the boom extension is offset you must _____.
    a. lower the main boom until the head sheave on the boom extension is touching the board on the ground
    b. remove the hoist rope from all sheaves
    c. release the lifting limit switch lock on the main boom
    d. spool the cable

24. In the power-pinned fly covered in this module, the section of the boom that is powered to extend or retract the fly is the _____.
    a. first section
    b. second section
    c. third section
    d. base section

25. When attaching an auxiliary single-sheave boom head to the main boom, the _____ need to be removed before laying the hoist rope for the auxiliary single-sheave boom head.
    a. mast attachments
    b. sheave retaining pins
    c. sheave attachment pins
    d. rope guide bars

GLOSSARY

Trade Terms Introduced in This Module

Auxiliary: As used in the text, a secondary hoist rope system used either in conjunction with, or independently of, the main hoist system.

Becket: A wedge socket that is used as an end preparation for wire rope.

Bushing: A replaceable hardened steel tube mounted in a case or housing as a bearing.

Idler: As used in the text, a sheave used to equalize tension in opposite parts of a rope.

Intermediate: Being situated between two points.

Lug: A projecting piece by which something is held or supported.

Mast: A frame hinged at or near the boom point for use in supporting a jib.

Pin: A piece of metal used to fasten, support, or attach components or members.

Turnbuckle: A link or sleeve used as a means of coupling and tightening two parts.

Figure Credits

Link-Belt Construction Equipment Company	202F01, 202F02
Topaz Publications, Inc.	202F03
Manitowoc Crane Group	202F04–202F11, 202RQ01, 202E01
PPM Terex	202F12–202F15,
Tony Guidry	202F16
Wire Rope Technical Board	202F17

CONTREN® LEARNING SERIES — USER UPDATE

The NCCER makes every effort to keep these textbooks up-to-date and free of technical errors. We appreciate your help in this process. If you have an idea for improving this textbook, or if you find an error, a typographical mistake, or an inaccuracy in NCCER's Contren® textbooks, please write us, using this form or a photocopy. Be sure to include the exact module number, page number, a detailed description, and the correction, if applicable. Your input will be brought to the attention of the Technical Review Committee. Thank you for your assistance.

Instructors – If you found that additional materials were necessary in order to teach this module effectively, please let us know so that we may include them in the Equipment/Materials list in the Annotated Instructor's Guide.

Write: Product Development and Revision
National Center for Construction Education and Research
P.O. Box 141104, Gainesville, FL 32614-1104

Fax: 352-334-0932

E-mail: curriculum@nccer.org

Craft ______ Module Name ______

Copyright Date ______ Module Number ______ Page Number(s) ______

Description ______

______

______

______

(Optional) Correction ______

______

______

(Optional) Your Name and Address ______

______

______

Module 38203-06

Lattice Boom Assembly and Disassembly

COURSE MAP

This course map shows all of the modules in the second part of the *Rigging* curriculum, *Rigging*. The suggested training order begins at the bottom and proceeds up. Skill levels increase as you advance on the course map. The local Training Program Sponsor may adjust the training order.

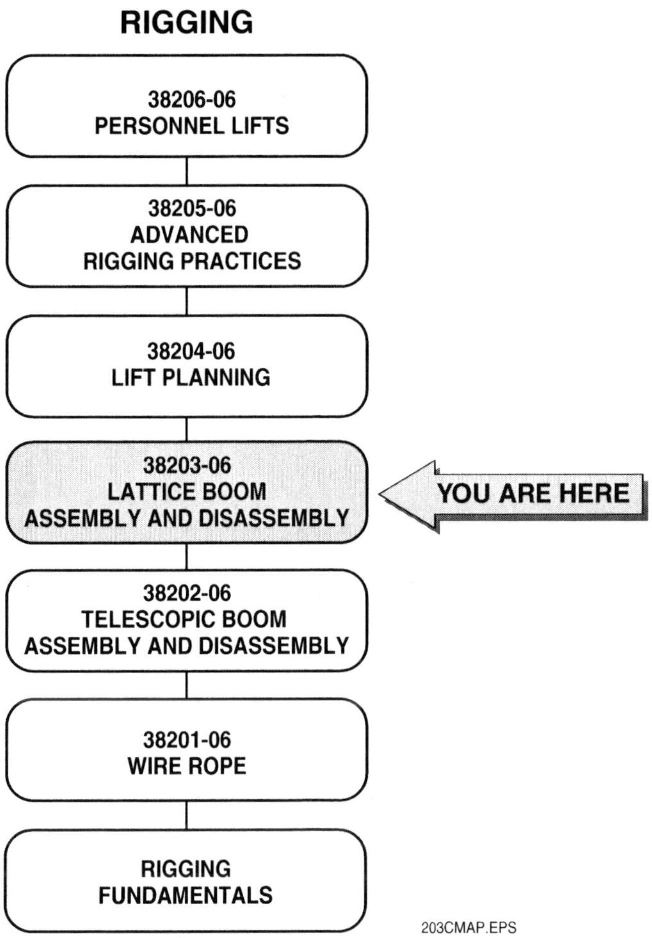

203CMAP.EPS

MODULE 38203-06 CONTENTS

1.0.0 INTRODUCTION 3.1
2.0.0 PRE- AND POST-ASSEMBLY CONSIDERATIONS 3.4
3.0.0 COUNTERWEIGHT CONSIDERATIONS 3.4
4.0.0 BOOM PARTS 3.5
5.0.0 LATTICE BOOM ASSEMBLY 3.5
5.1.0 Assembling Short Lattice Booms 3.6
5.2.0 Assembling Long Lattice Booms 3.9
6.0.0 JIB ASSEMBLY 3.12
7.0.0 DISASSEMBLY OF LATTICE BOOMS 3.12
SUMMARY 3.13
REVIEW QUESTIONS 3.13
GLOSSARY 3.17
ACKNOWLEDGMENTS 3.19

Figures

Figure 1 Crane assembly 3.2
Figure 2 Crane assembly procedures 3.3
Figure 3 Barricaded crane assembly area 3.4
Figure 4 Install counterweights 3.4
Figure 5 Pin secured with cotter pins .3.5
Figure 6 Welded lifting lugs 3.5
Figure 7 Extend the gantry 3.6
Figure 8 Install the boom base 3.7
Figure 9 Boom identification plate 3.7
Figure 10 Connect top pins 3.7
Figure 11 Install lower pins 3.8
Figure 12 Lower the boom onto the blocking 3.8
Figure 13 Cantilever length 3.9
Figure 14 Install counterweights 3.9
Figure 15 Blocking the front idlers on crawler ends 3.9
Figure 16 Secure lower pins 3.10
Figure 17 Pendant spreader bar 3.10
Figure 18 Move the pendants to the end of the boom section 3.10
Figure 19 Lower the boom onto blocking 3.11
Figure 20 Move the pendant to the end of the boom 3.11
Figure 21 Pre-raising checklist chart .3.11
Figure 22 Boom tip with jib bracket 3.12
Figure 23 Jib strut to jib base 3.12
Figure 24 Jib forestay and backstay pendants 3.12

MODULE 38203-06

Lattice Boom Assembly and Disassembly

Objectives

When you have completed this module, you will be able to do the following:

1. Determine if there is adequate space and resources for crane assembly and disassembly.
2. Identify lattice boom components.
3. Define the relationship of the counterweight to the assembly and disassembly of the boom.
4. Assemble and disassemble a short lattice boom.
5. Assemble and disassemble a long lattice boom.
6. Assemble and disassemble a jib at a lattice boom top.
7. Define and evaluate foundation requirements unique to long lattice boom erection.

Prerequisites

Before you begin this module, it is recommended that you successfully complete *Rigging Fundamentals*; and *Rigging*, Modules 38201-06 and 38202-06.

Required Trainee Materials

1. Pencil and paper
2. Appropriate personal protective equipment

1.0.0 ◆ INTRODUCTION

Depending on the crane and the overall weight, cranes must be disassembled for transportation and reassembled on site. For most lattice boom cranes, the boom and counterweights are removed. For larger cranes the crawler treads are also removed. Once the crane arrives at the site, it must be properly unloaded and assembled (*Figure 1*). After the lifts are completed, it must be disassembled and loaded for transportation.

Over half of all mobile crane accidents are the result of unskilled work performed when the crane is assembled and placed to make lifts. Operators and riggers need to ensure that the crane is assembled and disassembled correctly. The project engineer, site superintendent, contractor, or whoever is causing the crane to be used are also responsible for safety.

Specific guidelines and instructions for assembling the crane and installing various boom attachments and components are contained in the operator's safety manual for each crane. Some manufacturers provide a separate assembly manual.

A typical set of assembly procedures is shown in *Figure 2*. This module will give an overview of the assembly and disassembly procedures relevant to lattice boom cranes.

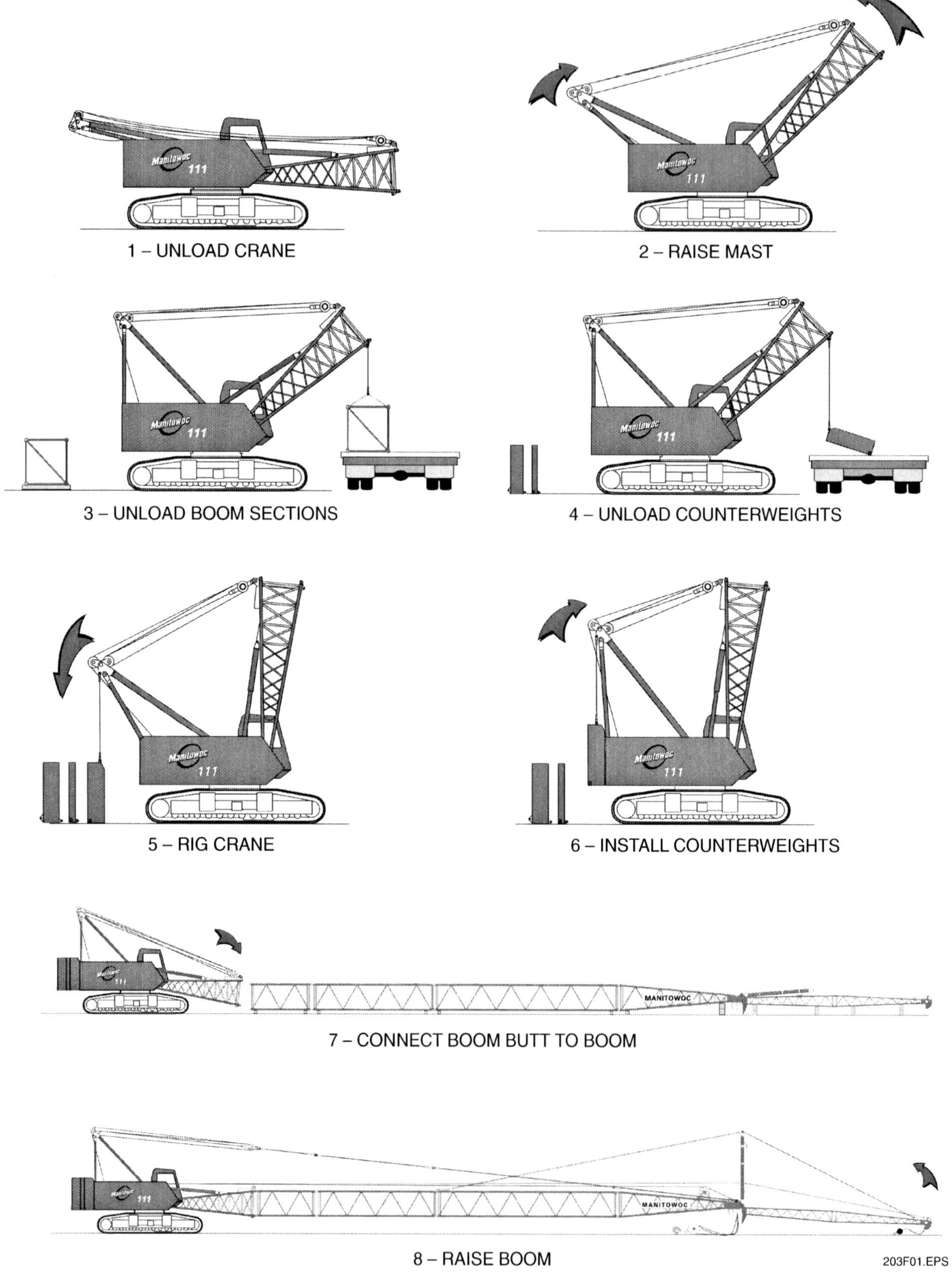

Figure 1 ◆ Crane assembly.

CRANE ASSEMBLY AND DISASSEMBLY GUIDE

MODEL 999

Table of Contents

Safety 2
Operating Limitations (Table 1) 2
General 3
Crane Weights 3
Operating Controls 3
Assembly Personnel 3
Assembly/Disassembly Area 3
Crane Orientation 3
Retaining Connecting Pins 3
Pre-Start Checks 3
Engine 3
Planetaries/Gear Cases 3
Hydraulic System 3

Assembly Procedure
A — Remove Crane from Trailer 4
B — Raise Mast To Operating Position 6
C — Install Assembly Block 8
D — Install Crawlers 11
E — Install Counterweight 14
F — Assemble Boom 18
G — Remove Boom Butt From Trailer 18
H — Remove Assembly Block 18
I-1 — Install #22E or #22EL Boom Butt 20
I-2 — Install #82 Boom Butt 22
J-1 — Connect #22E or #22EL Boom Butt To Boom 26
J-2 — Connect #82 Boom Butt To Boom 28
Pre-Raising Checks 31

Disassembly Procedure
A — Disconnect Boom Butt From Boom 31
General 31
Disconnect #22E or #22EL Boom Butt From Boom 32
Disconnect #82 Boom Butt From Boom 34
B-1 — Remove #22E or #22EL Boom Butt 36
B-2 — Remove #82 Boom Butt 38
C — Install Assembly Block 42
D — Install Boom Butt On Trailer 42
E — Disassemble Boom 42
F — Remove Counterweight 42
G — Remove Crawlers 44
H — Remove Assembly Block 48
I — Lower Mast To Transport Position 48
J — Install Crane On Trailer 48

Component Identification

NOTE: Except for procedures specific to #82 boom, mast for #22E boom is shown throughout this folio. Mast for #22EL and #82 booms is identical except that pendants (#22EL boom) or straps (#82 and #82LR boom) are suspended from mast.

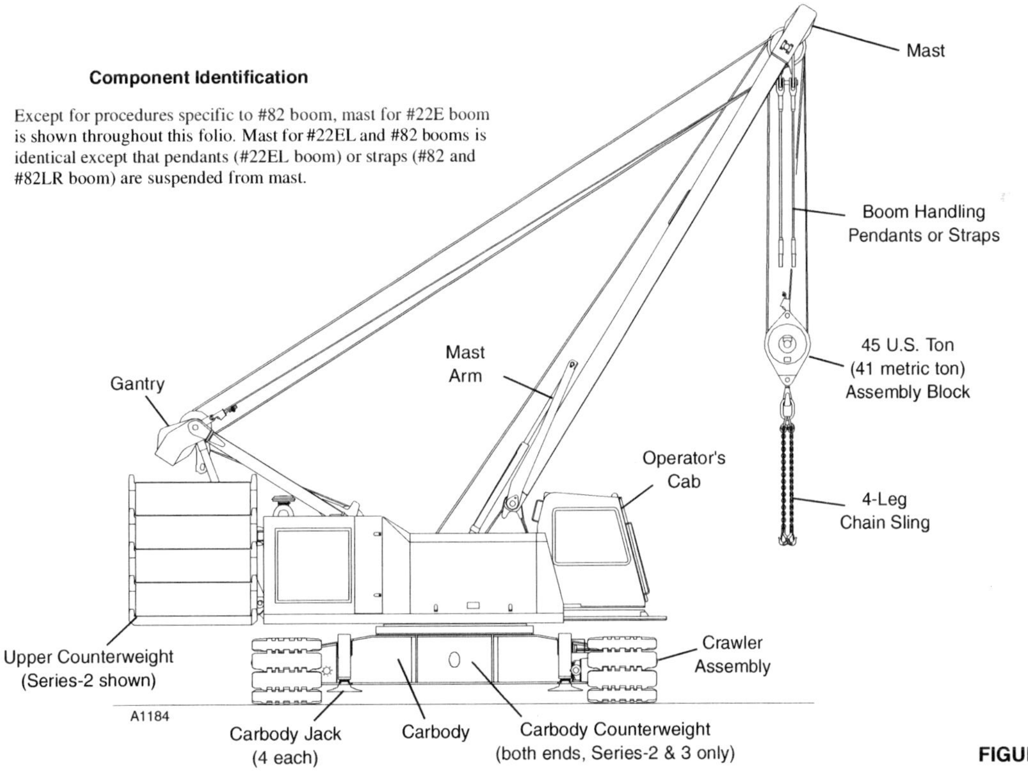

FIGURE 1

Rev. 02-13-2002 **Folio 1864-1**

203F02.EPS

Figure 2 ◆ Crane assembly procedures.

2.0.0 ◆ PRE- AND POST-ASSEMBLY CONSIDERATIONS

The first consideration for crane assembly is to select a good location to assemble the crane. The assembly/disassembly area should have a firm, level, uniformly supporting surface. Make sure that the area is large enough to accommodate the crane, the selected boom and jib lengths, and the movement of the trailer. An erection area should be barricaded two feet beyond the swing radius, as shown in *Figure 3*. It should be large enough to allow the components to be stacked and handled. It should also allow a crane to be assembled without interfering with or posing a hazard to site personnel. As the crane swings, there should be at least 2 feet of clearance between the counterweight and the nearest obstacle. The operator and the person responsible for the lift must consider the following:

- Is there enough room for the crane to maneuver into position on the site?
- Is there sufficient room in which to erect the boom?
- Can the trucks hauling boom sections get into place and be unloaded safely?
- Is there enough timber blocking to support the boom when it is being assembled and disassembled?
- Is there a good system for the crane operator to communicate with the assemblers?

> **WARNING!**
> Assembly and disassembly personnel must always wear the appropriate personal protective equipment and observe safe working practices when working with lattice boom sections.

Figure 3 ◆ Barricaded crane assembly area.

Figure 4 ◆ Install counterweights.

3.0.0 ◆ COUNTERWEIGHT CONSIDERATIONS

Crane operators and any personnel involved in the assembly and disassembly of lattice boom cranes must be aware of counterweight requirements. Do not begin assembling the boom until you find out what the correct counterweight is, and that it is sufficient to get the boom off the ground.

Never add or remove a counterweight from a carrier-mounted crane without first locating the machine on firm, level ground and extending all outrigger beams and jacks, as shown in *Figure 4*. In order to lift some long booms off the ground, it may be necessary to install extra counterweights beyond those which are required for normal operation. On some models of cranes, this extra weight must be removed before operations begin or when shorter booms are used. If it is not, the crane's backward stability will be reduced.

The swing lock must be engaged whenever any work is being performed on the counterweights and whenever the boom is being assembled, lifted, or lowered. Moving counterweights can cause the wire rope to break or the mast to collapse if the load exceeds the manufacturer's specifications.

> **WARNING!**
> To avoid tipping the crane over, install counterweights in the exact sequence given in the operator's manual. Lifting lugs may fail if the load exceeds the manufacturer's specifications. Falling counterweights can cause death, serious injury, or property damage.

4.0.0 ◆ BOOM PARTS

After the boom sections are unloaded, they should be inspected and inventoried. The boom sections should be inspected for damage or wear. Damaged or worn parts can cause serious accidents or injuries if they are used. They should be repaired or replaced pursuant to manufacturer's specifications.

Check the boom sections for bent **lacings**, bent or damaged main **chords**, or any broken or cracked **welds**. If the fittings are bent or damaged, the boom sections cannot be fit together easily. Check that the suspension ropes, **pendants**, and boom pendant pins are all in good condition. Pins must be retained by cotter pins, as shown in *Figure 5*.

Make sure that you have the correct size and quantity of nuts, bolts, and pins needed to secure the boom sections. Use only the fasteners specified in the operator's manual.

Before putting the boom together, determine the weight of the loads and rigging, as well as the height of the lift and the radius of operation. Find the shortest length that will accomplish the job. Check the operator's manual to ensure that the load weight and radius are within the tolerances listed.

Handle boom sections with care to prevent damage to chords, lacing, and connectors. It is very important to pay close attention to the type of sling being used for assembly. Slings made of nylon are best for handling boom sections, although wire rope or chain slings can be used. When using wire or chain slings, stay clear of lacing to prevent damage. Some boom sections have welded lifting lugs as shown in *Figure 6*. If the boom section has lifting lugs, they should be used for lifting the boom sections.

Figure 5 ◆ Pin secured with cotter pins.

Figure 6 ◆ Welded lifting lugs.

5.0.0 ◆ LATTICE BOOM ASSEMBLY

The information in this module is presented as a general guideline. Always refer to the manufacturer's manual for the specific lattice boom crane before any assembly work is done.

A boom rigging drawing is furnished with each machine for the particular boom ordered. Important rigging information, including boom assembly, **boom gantry** and back hitch assembly, boom point assemblies, and pendants, are contained within the rigging drawing. Assembly personnel should read, study, and understand all the content before assembly.

> **WARNING!**
> Boom section connecting points can be hazardous. Do not perform maintenance directly under a boom. Collapse could occur, resulting in serious injury or death.

5.1.0 Assembling Short Lattice Booms

This section presents a general guideline for assembling short lattice booms. Always consult the manufacturer's recommendations before any assembly or disassembly takes place.

Step 1 Verify the amount of counterweight required in the load/capacity chart for the lifts to be made and the length of boom being installed. Make sure enough counterweight is installed to lift the boom off the ground.

Step 2 If so equipped, the extendable counterweight must be properly extended to manufacturer requirements.

Step 3 Rotate the upperworks to face the direction recommended by the manufacturer, or in the direction of maximum stability.

Step 4 Set the swing lock.

Step 5 Fully extend the gantry (see *Figure 7*). If the machine has a live **boom mast**, check the load/capacity chart to see if it must be used. Never attempt to raise a boom until this equipment has been erected. Crane manufacturers will specify the maximum length of boom which can be raised with standard equipment using the crane machinery, and which lengths require the use of special equipment.

Step 6 Install the boom base (*Figure 8*) section, and attach the supporting pendants.

WARNING!
Boom sections must be installed using the sequence and methods specified by the manufacturer. A boom can collapse if not properly assembled. Do not mix and match boom sections.

Step 7 Assemble the boom sections in the order specified by the manufacturer's manual. The boom number may be stamped on the side of the boom joint connector on two diagonal chords of each boom section. The boom may also have an identification plate, as shown in *Figure 9*. Boom identification is done to prevent mixing boom sections of one type with those of another. Boom identification numbers can exist as either a boom number (20), a boom component part number (42597-2), or a date-manufactured number (12-10-92).

WARNING!
Do not stand inside the boom while installing connector pins. Stand outside the boom.

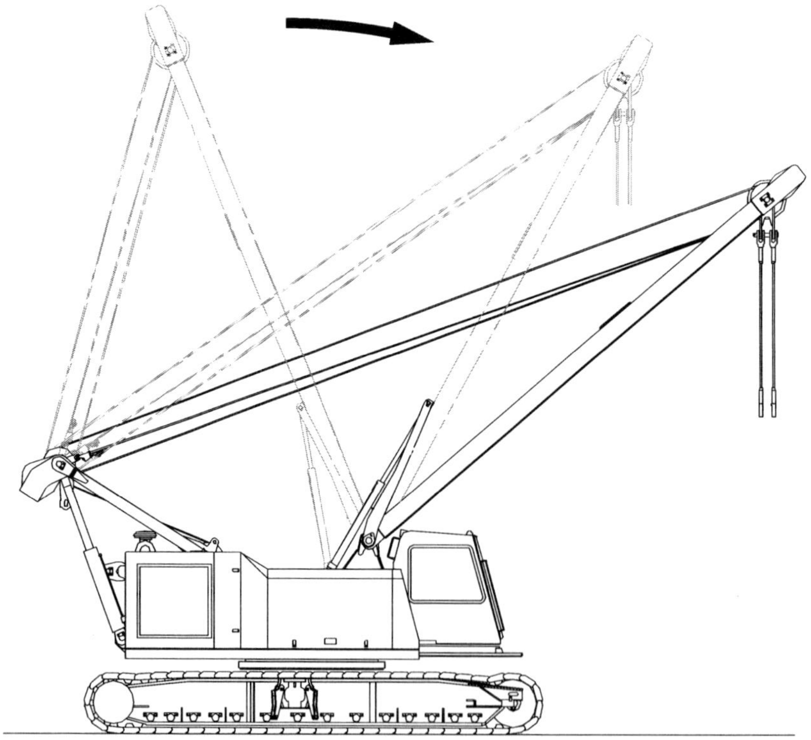

203F07.EPS

Figure 7 ◆ Extend the gantry.

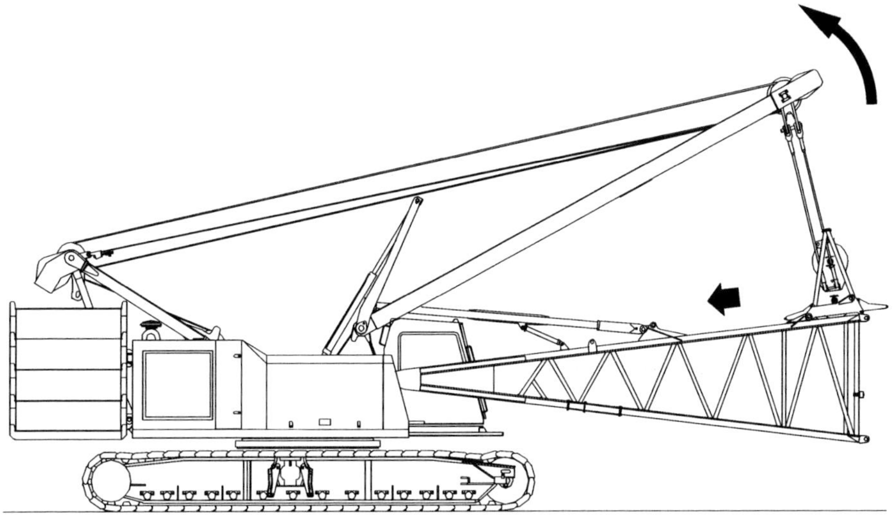

Figure 8 ◆ Install the boom base.

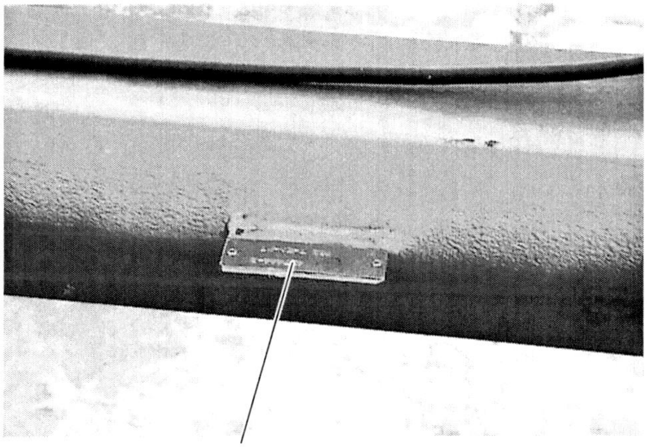

Figure 9 ◆ Boom identification plate.

Step 8 Insert the top pins first.

> **WARNING!**
> Ensure that all personnel are clear before operating the crane.

Step 9 Keep the pendants attached behind the boom sections being connected. Line up the two upper pin connection holes between the two sections being pinned together. Insert the two top boom pins and install the pin retainers (*Figure 10*).

Step 10 Lay out the remaining sections, draw them together, and insert all top pins.

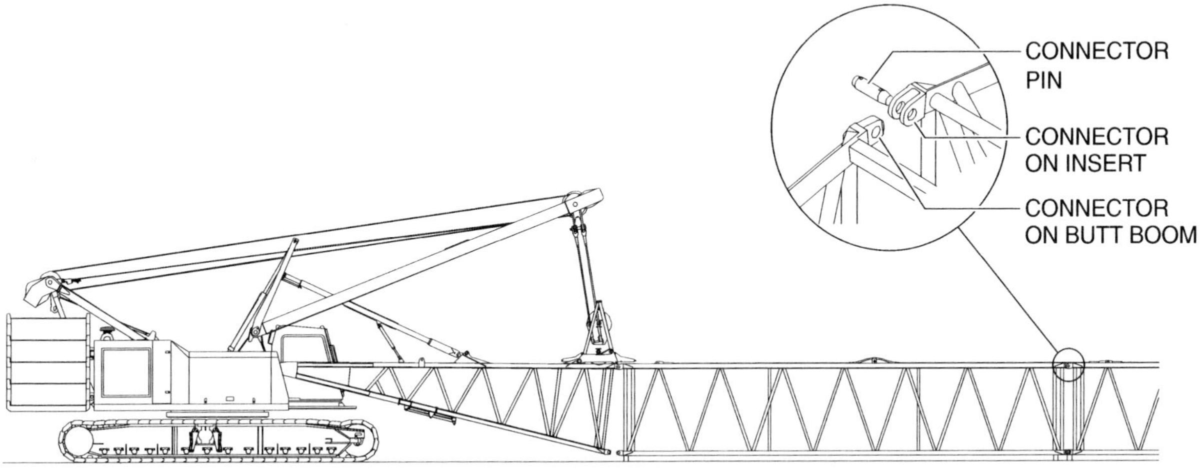

Figure 10 ◆ Connect top pins.

> **CAUTION**
> When the boom is assembled, check that the pendants are the proper length for the boom length.

Step 11 If the pendants are slightly long on a machine with a short boom, the weight of the live mast will pull an unloaded boom into the backstops when the minimum radius is approached. The boom will have to be pulled away from the backstops by hanging a load on the hook.

> **CAUTION**
> Make sure that the suspension ropes and pendants do not interfere with the boom pins and/or clips that maintain the pin connections.

> **WARNING!**
> Check the load capacity charts before raising the boom. It may be necessary to block the crawlers to maintain stability.
>
> It is very important to have the pendants exactly the same length. If they aren't, the boom will be twisted, and the load will be altered. This creates the potential for structural failure.

Step 12 Boom up slightly until the bottom pin connection points line up. Engage the boom hoist pawl. Install the pins and retainers (*Figure 11*).

Step 13 Lower the boom onto blocking, and extend the pendants out to the boom tip (*Figure 12*). Pendants come in matched sets of either two or four. Identification numbers are stamped into both ends of each pendant to aid in keeping them in sets. Always install pendants in sequence as noted on the rigging drawing for the particular boom length being used.

Figure 12 ◆ Lower the boom onto the blocking.

Step 14 Rig the main hoist line (and auxiliary or whip line, if necessary).

Step 15 Check that the boom angle indicator is properly installed and adjusted.

Step 16 Boom up slightly, engage the boom hoist pawl, and reeve the block and/or ball.

Step 17 Make sure that the boom hoist wire rope is spooled tightly onto the drum and engaged with the proper sheaves, and that the wire rope is securely anchored to the wedge socket at the gantry.

Step 18 Do not lift the load block or ball when raising the boom. Pay out the hoist line to keep the block on the ground. Raise the boom smoothly in one continuous lift until the minimum boom angle is reached.

> **CAUTION**
> Always follow manufacturer recommendations when raising the boom to avoid damaging the equipment.

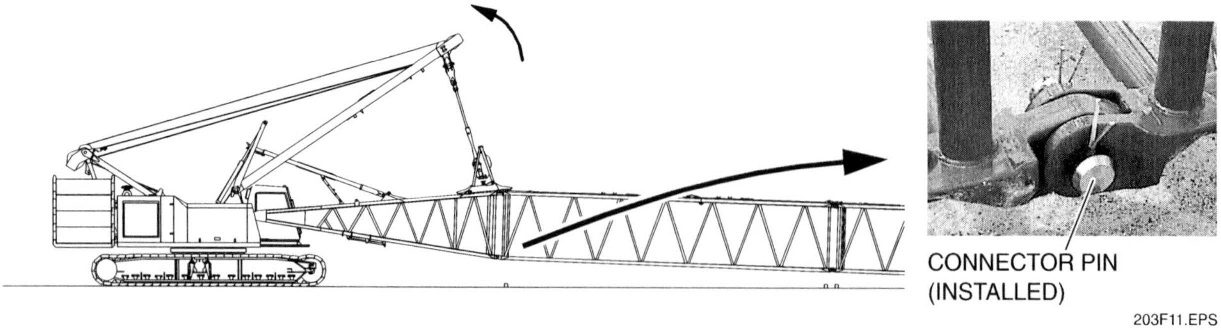

Figure 11 ◆ Install lower pins.

Step 19 Once the boom is in the air, check that the boom hoist limiting device (kickout) is installed and in working order.

5.2.0 Assembling Long Lattice Booms

This assembly method applies when the boom length exceeds the manufacturer's maximum **cantilever** length (*Figure 13*). As with short lattice booms, this is a guideline, and any manufacturer's recommendations must be followed.

Step 1 Check the amount of counterweight required in the load/capacity chart for the lifts to be made and the length of boom being installed. Check to ensure that enough counterweight is installed to lift the boom off the ground (*Figure 14*).

Step 2 Check to see if a front bumper counterweight is required.

Step 3 If so equipped, extend the extendible counterweight.

WARNING!
The appropriate counterweights must be used, or the crane may tip.

Step 4 Rotate the upperworks to face the direction recommended by the manufacturer or the direction of maximum stability.

Step 5 Set the swing lock.

Step 6 Fully extend the gantry. If the machine has a live mast, check the load/capacity chart to see if it must be used. Never attempt to raise a boom until this equipment has been fitted. Crane manufacturers will specify the maximum length of boom which can be raised with standard equipment using the crane machinery and which lengths require the use of special equipment.

Step 7 Certain crane manufacturers recommend blocking the front idlers on the crawler ends (*Figure 15*). Consult the operator's manual.

Step 8 Install the base or foot section of the boom, and attach the supporting pendants to the ends.

Step 9 Assemble the boom base sections in the order specified by the manufacturer.

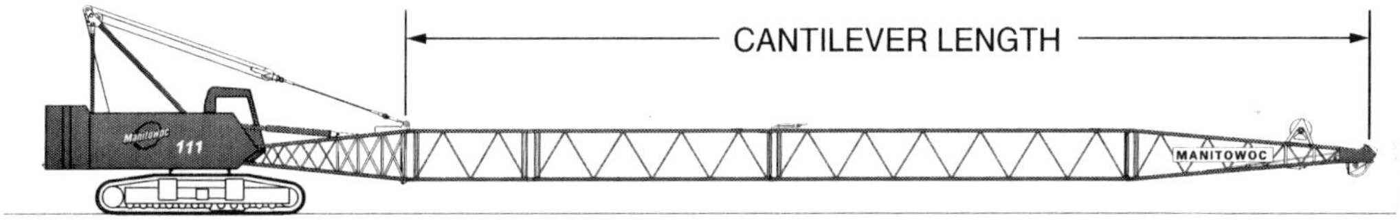

203F13.EPS

Figure 13 ◆ Cantilever length.

203F14.EPS

Figure 14 ◆ Install counterweights.

203F15.EPS

Figure 15 ◆ Blocking the front idlers on crawler ends.

Step 10 Draw some, but not all, of the sections together. The total length of the inserts to assemble at this stage must not exceed the maximum cantilever length specified by the manufacturer. Line up the upper pin connection points. Insert the top pins only, and secure them with pin retainers.

Step 11 Boom up slightly until the bottom pin connection points line up. Engage the boom hoist pawl. Install the pins and pin retainers (*Figure 16*).

Step 12 Lower the boom onto blocking until the pendants are slack. Use pendant **spreader bars** (*Figure 17*) as required by the manufacturer.

Step 13 Move the pendants out to the end of the last insert (*Figure 18*). Remember to watch for twisting of the boom.

WARNING!
If the pendants are too short on a machine with a longer boom, it will overload the boom, mast, pendants, and boom hoist systems when handling near-capacity loads.

Step 14 Draw the remaining boom sections together. Ensure that their length does not go over the maximum cantilever length.

Step 15 Line up the upper pin connection points. Insert the top pins only, and secure them with retainer pins.

Step 16 Boom up slightly until the bottom pin connection points line up. Engage the boom hoist pawl. Install pins and cotters.

Step 17 Lower the boom onto blocking until the pendants are slack (*Figure 19*).

203F16.EPS

Figure 16 ◆ Secure lower pins.

203F17.EPS

Figure 17 ◆ Pendant spreader bar.

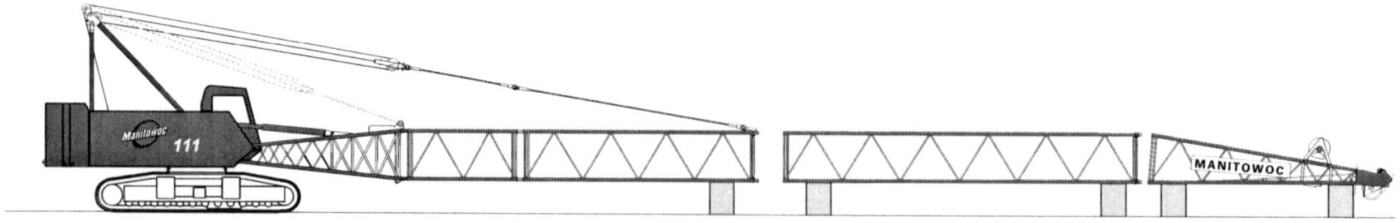

203F18.EPS

Figure 18 ◆ Move the pendants to the end of the boom section.

203F19.EPS

Figure 19 ◆ Lower the boom onto blocking.

Step 18 Move the pendants out to the end of the boom (*Figure 20*).

Step 19 Rig the main hoist line, and if necessary, the whip line.

> **NOTE**
>
> Some manufacturers require boom suspension for long booms.

Step 20 Boom up slightly, engage the boom hoist pawl, and reeve the block.

Step 21 To gain extra stability, do not lift the load block when raising the boom. Pay out the hoist line to keep the block on the ground. Raise the boom slowly.

Some manufacturers recommend certain checks be performed before the boom is raised. An example of these checks is shown in *Figure 21*.

Pre-Raising Checks

Make following checks and correct any defects before raising attachment:

- ❑ Crane is on firm, level surface.
- ❑ Boom hinge pins are fully engaged and locking cover is pinned in up position.
- ❑ Crawler connecting pins are engaged and locking pins are installed.
- ❑ Carbody jack pads are stored.
- ❑ Carbody jacks are fully retracted and pinned in stored position.
- ❑ Boom and jib inserts are installed in proper sequence per rigging drawing
- ❑ Spreader (22EL boom only) is installed at proper location and spread to proper position.
- ❑ Intermediate suspension installed if required (#82 Long Reach boom only).
- ❑ Links properly connected between all straps.
- ❑ All insert and pendant connector pins installed. Cotter pins installed and spread.
- ❑ Gantry fully raised.
- ❑ Mast arms fully lowered.
- ❑ Mast arms bypass handle turned off.
- ❑ Setup mode turned off (desired operating mode is selected and confirmed).
- ❑ Boom hoist wire rope spooled tightly onto drum and engaged with proper sheaves. Wire rope securely anchored to wedge socket at gantry.
- ❑ Load lines spooled tightly onto drums and engaged with proper sheaves. Load lines securely anchored to wedge sockets at boom and jib point or at load block and weight ball.
- ❑ All blocking, tools, and other items removed from boom and jib.
- ❑ Automatic boom stop properly installed and adjusted (refer to Boom Stop Adjustment folio in Operator's Manual).
- ❑ Cable from crane control system connected to cable reel on boom butt.
- ❑ Block-up limit control properly installed and operational.
- ❑ LMI properly installed and operational (refer to Block-Up Limit Adjustment folio in Operator's Manual).
- ❑ Crane and attachment properly lubricated (refer to Lubrication Guide in your Operator's Manual).
- ❑ Crawlers blocked if required per capacity chart.
- ❑ Wind within allowable limits for operation given on capacity chart.

203F21.EPS

Figure 21 ◆ Pre-raising checklist chart.

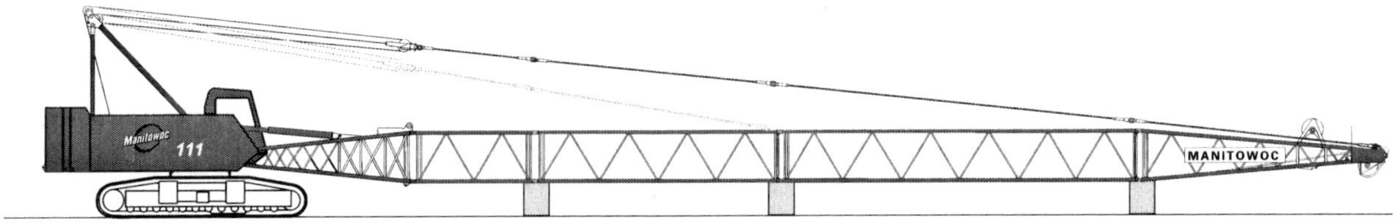

203F20.EPS

Figure 20 ◆ Move the pendant to the end of the boom.

6.0.0 ◆ JIB ASSEMBLY

Once the boom is assembled, the jib can be assembled and attached to the boom. The following steps are general guidelines for installing a jib onto a lattice boom tip:

Step 1 Lay the boom tip on blocking on the ground. Some boom tips have welded brackets to attach the jib, as shown in *Figure 22*. If a jib adapter is required for the boom, install the jib adapter to the boom tip.

Step 2 Assemble the jib on blocking on the ground, pinning the jib base to the boom tip.

Step 3 Pin the jib strut to the jib base (*Figure 23*).

Step 4 Pin the jib forestay and backstay pendants to the strut tip with the pins furnished with the pendants (*Figure 24*).

Step 5 Install a wire-rope roller guide on each jib insert used, if applicable.

Step 6 Pin the other end of the jib pendants to the jib point.

203F22.EPS

Figure 22 ◆ Boom tip with jib bracket.

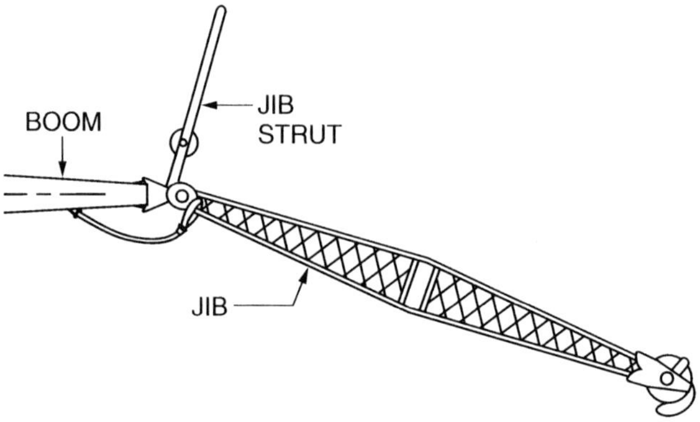

203F23.EPS

Figure 23 ◆ Jib strut to jib base.

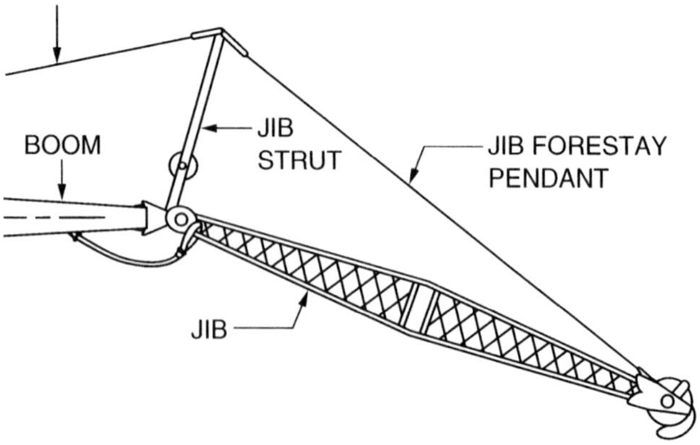

203F24.EPS

Figure 24 ◆ Jib forestay and backstay pendants.

7.0.0 ◆ DISASSEMBLY OF LATTICE BOOMS

The following sequence is a guideline for removing lattice boom sections. Always consult manufacturer requirements when performing any kind of disassembly.

WARNING!
Boom section connecting points can be hazardous. Do not perform maintenance directly under a boom. Collapse could occur, resulting in serious injury or death.

Step 1 Move the crane onto a level area that is cleared of personnel.

Step 2 Set the outriggers and level the machine, or block idlers on a crawler.

Step 3 Check that the counterweight is correct.

Step 4 Set the swing lock.

CAUTION

A thorough inspection of the boom hoist cable should be made before booming down. Cable damaged during crane operations can break under maximum load. Collapse could occur, resulting in serious equipment damage.

Step 5 Lower the boom over the rear or over the area of maximum stability.

Step 6 Place blocking under the boom tip and under both sides of every joint, if possible.

Step 7 Lower the boom onto this blocking until the pendants go slack.

Step 8 Move the pendants back by removing a length and re-anchoring them at a lower boom section.

Step 9 Elevate the boom slightly. Do not elevate it so far that the boom peak is lifted off its blocking.

Step 10 Drive out the lower pins on the sections ahead of the pendants.

Step 11 Lower the boom back down onto the blocking. Make sure every section is supported.

Step 12 Drive out the upper pins on the sections ahead of the pendants.

Step 13 Repeat this procedure until the boom is disassembled.

Summary

Lifting operations depend on proper assembly of the crane's boom. On lattice boom cranes, this involves the proper attachment of the boom sections. It is imperative that assembly and disassembly personnel be aware of clearance requirements on the job site, and that they also follow the appropriate safety requirements associated with the work.

Review Questions

1. For each crane, guidelines and instructions for installing various boom attachments and components are _____.
   a. contained in the operator's safety manual
   b. listed on the manufacturer's web site
   c. the same regardless of serial number, as long as the crane is manufactured by the same company
   d. the same regardless of model number, as long as the crane is manufactured by the same company

2. Cranes must be assembled on site because _____.
   a. the configuration depends on site conditions
   b. each part must be separately inspected
   c. new wire ropes must be installed
   d. they are too heavy to transport intact

3. Over half of all mobile crane accidents are the result of _____.
   a. poorly-trained operators
   b. mechanical and equipment failures
   c. unskilled work performed when the crane is assembled
   d. boom failure due to overweight load

4. Disassembly of most lattice boom cranes involves removal of the _____.
   a. carbody and counterweights
   b. outriggers and boom
   c. base mounting and counterweights
   d. boom and counterweights

5. The first concern for crane assembly is to _____.
   a. select a good location
   b. make sure you have all the pins and bolts
   c. calculate the counterweight needed
   d. inspect all the boom sections

6. The crane assembly area must be _____.
   a. close to or at the lift location
   b. on a hill to make it easier to lift the boom
   c. a firm and level surface
   d. close to the main access road

7. Barricades should be placed _____.
   a. completely around the assembly area
   b. at a distance of 5 feet around the crane body
   c. at the swing radius of the crane
   d. around boom sections

8. Assembly personnel should check if there is enough timber to _____.
   a. chock the wheels of delivery trucks
   b. brace the crawlers on the crane
   c. block the boom sections
   d. build barricades

9. The crane operator and the assemblers must have _____.
   a. the same instruction book
   b. a way to communicate
   c. crane operation training
   d. commercial driver's licenses

10. In order to lift a longer boom off the ground it may be necessary to install _____ on the crane.
   a. a boom extension tip
   b. pendant spreaders
   c. a luffing jib
   d. additional counterweights

11. Whenever any work is being performed on the counterweights, the ____ should be engaged.
   a. swing lock
   b. load moment indicator
   c. anti-two-blocking device
   d. emergency brake

12. Before adding or removing counterweights you must _____.
   a. locate the crane on a firm level surface
   b. retract the outriggers
   c. chock the crawlers
   d. lockout forward motion control

13. Before adding or removing counterweight from a carrier-mounted crane, the _____ must be extended.
   a. hoist rope
   b. telescopic boom sections
   c. pendant spreader bars
   d. outrigger beams and jacks

14. If the fittings on boom sections are bent, _____.
   a. the pins should be replaced
   b. use larger pins to fill the gap
   c. the boom sections will not fit together properly
   d. do not use them to lift the boom

15. After the boom parts are unloaded they should be _____.
   a. placed for assembly
   b. properly lubricated
   c. connected with pins
   d. inspected for damage

16. All pins should be secured with _____.
   a. cotter pins
   b. wire rope
   c. bolts
   d. nuts

17. The boom sections should be inspected for _____ before assembly.
   a. proper length
   b. correct order
   c. bent lacings
   d. manufacturer's identification

18. Before assembly, suspension ropes and pendants should be in good condition and _____.
   a. well lubricated
   b. reeved into the block
   c. coiled on the drum
   d. retained by cotter pins

19. When you determine the weight of the loads, the height of the lift, and the radius of operation, the boom length should be the shortest length that will accomplish the job.
   a. True
   b. False

20. When handling the boom, to prevent damage you should _____.
   a. always use chain slings
   b. use a choker hitch around the lacing
   c. attach the sling to the lifting lugs if present
   d. use a basket sling around boom section

21. Any pins and fasteners can be used to assemble the boom as long as they are secured by cotter pins.
   a. True
   b. False

22. A rope or bar of specified length with fixed end connections is called a _____.
   a. cantilever
   b. pendant
   c. placard bar
   d. jib adapter

23. Information on boom assembly, gantry and back hitch assembly, boom point assemblies, and pendants is available _____.
    a. in the manufacturer's inspection manual
    b. on the boom rigging drawing
    c. at the manufacturer's web site
    d. on the range diagram

24. Assembly personnel must make sure that the _____ interfere with the boom pins and/or clips that maintain the pin connections.
    a. suspension ropes do not
    b. anti-two-block device does not
    c. load cell does not
    d. gantry does not

25. Pendants may come in matched sets of _____.
    a. four
    b. six
    c. eight
    d. ten

26. After rigging the main hoist line during boom assembly, the assembly personnel must make sure the _____ is properly installed and adjusted.
    a. boom angle indicator
    b. boom length indicator
    c. hoist drum rotation indicator
    d. selected controls system

27. The amount of counterweight needed to lift the boom can be verified by _____.
    a. asking the project manager
    b. using the load chart
    c. using all the weights on hand
    d. checking the project specifications

28. The boom identification number is used _____.
    a. by the manufacturer for tracking
    b. to date-stamp each section
    c. only for inventory control
    d. to prevent mixing boom sections of different types

29. When connecting boom sections, the bottom pins should be secured first.
    a. True
    b. False

30. To aid in keeping pendant sets together, both ends of each pendant are _____.
    a. color coded
    b. stamped with identification numbers
    c. clipped together
    d. the same size

31. When raising the boom, the block must be _____.
    a. on the ground
    b. at the boom tip
    c. level with the boom base
    d. exactly half the distance to the ground

32. A frame hinged at or near the boom hinge for use in supporting a boom is called a boom _____.
    a. gantry
    b. pendant
    c. chord
    d. mast

33. When assembling long lattice booms, each group raised _____.
    a. cannot exceed 20 feet
    b. must be a minimum of two boom sections
    c. cannot exceed the maximum cantilever length
    d. are a maximum of three boom sections

34. A thorough inspection of _____ should be made before booming down.
    a. the hydraulic system
    b. outrigger positioning
    c. boom hoist cable
    d. the pendants

35. In the jib assembly procedure discussed in this module, the jib is assembled and attached to the lattice boom tip after the _____.
    a. boom is assembled
    b. wire rope roller guide is installed on each jib insert
    c. jib strut is pinned to the jib boom
    d. jib forestay is pinned to the strut tip

GLOSSARY

Trade Terms Introduced in This Module

Boom gantry: A structural frame, extending above the crane superstructure, to which the boom or boom mast support ropes are reeved.

Boom mast: A frame hinged at or near the boom hinge for use in supporting a boom. The head of the mast is usually supported and raised or lowered by the boom hoist ropes.

Cantilever: A rigid structural member projecting from a vertical support so that the upper part is in tension and the lower part is in compression.

Chord: A load-bearing member of a lattice boom.

Lacing: A lattice strut member of a boom structure that is used to add structural support to the boom and keep the chords in their proper position.

Pendant: A rope or bar of specified length with fixed end connections.

Spreader bars: Adjustable bars used with long boom lengths to keep the pendants from scuffing on top of the boom.

Weld: A localized body of metals or nonmetals produced either by heating the materials to the melting temperature, with or without the application of pressure; or by the application of pressure alone, and with or without the use of filler material.

Figure Credits

Manitowoc Crane Group	203F01, 203F02, 203F05–203F11, 203F13, 203F21
Topaz Publications, Inc.	203F03, 203F12, 203F16, 203F19
Link-Belt Construction Equipment Company	203F04, 203F14, 203F15, 203F17, 203F22

CONTREN® LEARNING SERIES — USER UPDATE

The NCCER makes every effort to keep these textbooks up-to-date and free of technical errors. We appreciate your help in this process. If you have an idea for improving this textbook, or if you find an error, a typographical mistake, or an inaccuracy in NCCER's Contren® textbooks, please write us, using this form or a photocopy. Be sure to include the exact module number, page number, a detailed description, and the correction, if applicable. Your input will be brought to the attention of the Technical Review Committee. Thank you for your assistance.

Instructors – If you found that additional materials were necessary in order to teach this module effectively, please let us know so that we may include them in the Equipment/Materials list in the Annotated Instructor's Guide.

Write: Product Development and Revision
National Center for Construction Education and Research
P.O. Box 141104, Gainesville, FL 32614-1104

Fax: 352-334-0932

E-mail: curriculum@nccer.org

Craft ______ Module Name ______

Copyright Date ______ Module Number ______ Page Number(s) ______

Description ______

(Optional) Correction ______

(Optional) Your Name and Address ______

Module 38204-06

Lift Planning

COURSE MAP

This course map shows all of the modules in the second part of the *Rigging* curriculum, *Rigging*. The suggested training order begins at the bottom and proceeds up. Skill levels increase as you advance on the course map. The local Training Program Sponsor may adjust the training order.

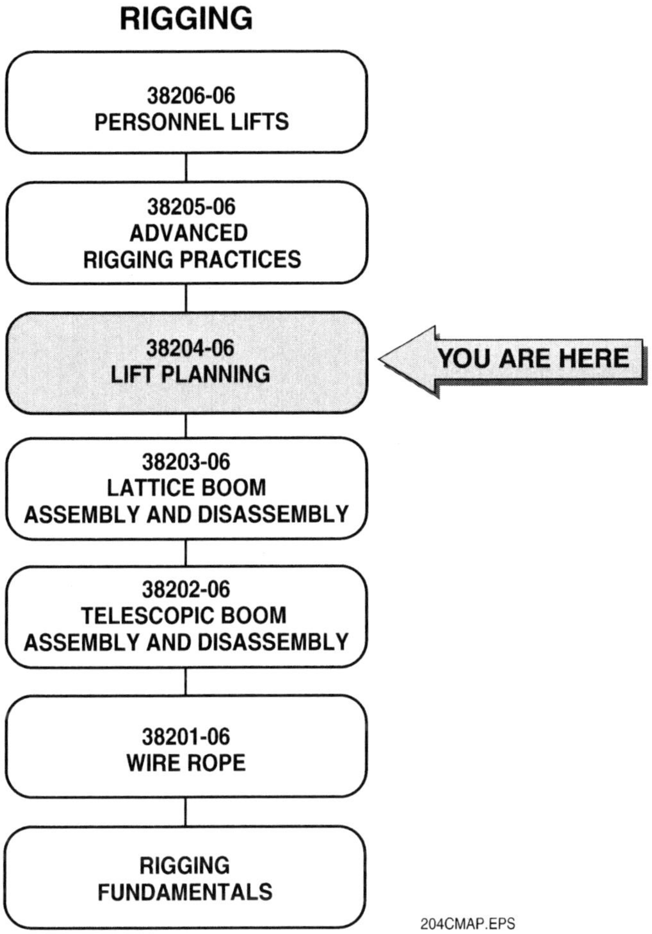

204CMAP.EPS

MODULE 38204-06 CONTENTS

1.0.0 INTRODUCTION 4.1

2.0.0 LIFT PLAN 4.1

2.1.0 Reference Information 4.2

3.0.0 CALCULATIONS 4.2

3.1.0 Single-Crane Lifting 4.2

3.2.0 Multiple-Crane Lifting 4.5

4.0.0 CRITICAL LIFTS 4.14

5.0.0 ENGINEERING CONSIDERATIONS 4.14

6.0.0 LIFT PLAN IMPLEMENTATION 4.15

SUMMARY 4.15

REVIEW QUESTIONS 4.15

GLOSSARY 4.19

APPENDIX, Critical Lift Plan 4.21

ACKNOWLEDGMENTS 4.35

Figures

Figure 1 Load/capacity chart one 4.3

Figure 2 Load/capacity chart two 4.4

Figure 3 Load/capacity chart three 4.5

Figure 4 Three-crane lift 4.6

Figure 5 Multiple-crane lift 4.7

Figure 6 Multiple-crane lift – center of gravity location 4.8

Figure 7 Multiple-crane lift – distance B, trial 1 .... 4.8

Figure 8 Multiple-crane lift – distance B, trial 1 results 4.9

Figure 9 Multiple-crane lift – distance B, trial 2 .... 4.9

Figure 10 Multiple-crane lift – distance B, trial 2 results 4.10

Figure 11 Multiple-crane lift with equalizer beam 4.11

Figure 12 Multiple-crane lift with equalizer beam, determining distance B and C 4.12

Figure 13 Multiple-crane lift with equalizer beam, distance B and C results 4.13

MODULE 38204-06

Lift Planning

Objectives

When you have completed this module, you will be able to do the following:

1. Provide the necessary information requested on a lift plan.
2. Reference available material that will assist in a safe lifting operation.
3. Calculate additions and deductions involved in lifting operations.
4. Identify existing operations that need special approval.
5. Identify engineering considerations in a lift plan.
6. Identify the various types of lift plans and their differences.
7. Identify the importance of lift plan implementation.
8. Describe the importance of following and adhering to a lift plan.

Prerequisites

Before you begin this module, it is recommended that you successfully complete *Rigging Fundamentals*; and *Rigging*, Modules 38201-06 through 38203-06.

Required Trainee Materials

1. Pencil and paper
2. Appropriate personal protective equipment
3. Calculator

1.0.0 ◆ INTRODUCTION

Before conducting a lift, most construction sites or companies require that a **lift plan** be completed and signed by competent personnel. Lift plans are mandatory for steel erection and multiple-crane lifts. A lift plan contains information relative to the crane, load, and rigging, and it lists any special instructions or restrictions for the lift. It is important to remember that any deviation from the original lift plan, no matter how small, requires review of the lift plan, and possibly a new lift plan.

2.0.0 ◆ LIFT PLAN

Critical lifts always require a lift plan. As a matter of policy, all operators should require or prepare a plan for all lifts. A typical lift plan for either a critical or ordinary (minor) lift breaks down the following elements and information (see the *Appendix*):

- Weight
  - Equipment condition
  - Weight empty
  - Weight of overhead ball
  - Weight of block
  - Weight of lifting bar
  - Weight of slings and shackles
  - Weight of jib (if erected or stowed and not used)
  - Weight of overhaul ball on jib
  - Weight of wire rope
  - Allowance for unaccounted material in equipment
- Jib
  - Jib to be used
  - Length of jib
  - Angle of jib
  - Rated capacity of jib
- Crane placement
  - Smooth solid foundation within one percent of grade
  - Electrical hazards in area
  - Obstacles or obstructions to lift or swing
  - Swing direction and degree
  - Area of operation (over-the-front, over-the-rear, over-the-side)

- Rope
  - Number of parts of rope
  - Size of rope
  - Reeving capacity
- Sizing and types of slings
  - Sling selection
  - Shackle selection
  - Sizing and types
- Crane
  - Type of crane
  - Crane capacity
  - Lifting arrangement
- Other
  - Special instructions or restrictions for crane, rigging, or lift
  - Diagrams for crane, load placement, and rigging configuration
  - Signature of job supervisor, lift plan checker, and operator
  - Derigging plan

2.1.0 Reference Information

The personnel involved in planning the lift should consult reference information that will assist in a safe lifting operation. This includes crane specifications, conditions in which the lift is taking place (weather, lift area space), and all applicable compliance and consensus standards. Standards covering mobile cranes include the following:

- *OSHA 29 CFR 1910.180*
- *OSHA 29 CFR 1926.550*
- *ASME B30.5*

3.0.0 ◆ CALCULATIONS

There are numerous calculations involved in lift planning. The personnel involved in planning the lift must take certain considerations into account:

- Available lifting-point height above the lifting location
- Load pick-up height above or below the lifting location
- Load set-down height above or below the lifting location
- Maximum wire rope weight to be experienced during the lift
- Weight of load to be lifted, not including rigging
- Weight of crane components below the lifting point (deductions)
  - Length of line from point to load block
  - Any parts of line that are considered part of the load
  - Load block height
  - Rigging on the block, including lifting beams, slings, and shackles
- Effective weight of crane components past the lifting point
  - Jib
  - Line to jib top
  - Length of line from point to load block
- Weight of any additional crane components and/or boom components

3.1.0 Single-Crane Lifting

The information provided is for instructional purposes only. The load/capacity charts presented here are derived from manufacturer sales information that may not be complete or machine-specific. They are intended to give you practice in lift calculations. On a job site, always follow the manufacturer's load charts and recommendations.

To illustrate lift plan calculations, consider this example. What is the net capacity of the Manitowoc 4100W-2 in the following configuration?

- 22C boom, 190-ft main boom, no offset
- 55 ft of load radius
- 2-sheave boom point attached
- 15-ton hook block 20 ft below boom point
- 100-ton block reeved with 4 parts of line
- Crawlers extended
- Rigging of 365 lbs
- Lifting from the main boom

Refer to the Manitowoc 4100W-2 load/capacity charts shown in *Figures 1* through *3*.

The gross capacity of the crane at 190 ft of boom length and 55 ft of load radius is 88,000 lbs. Deductions from this weight include:

- 2-sheave boom point = 1,500 lbs
- 15-ton hook block or hook and weight ball = 865 lbs
- 100-ton block = 2,065 lbs
- Rigging = 365 lbs
- Rope = 1,827 lbs

The total weight of the deductions is 6,622 lbs. The net capacity of the crane in this configuration is the gross capacity (88,000 lbs) minus the deductions (6,622 lbs), which is 81,378 lbs.

MANITOWOC 4100W-2

LIFTCRANE CAPACITIES — MEETS ANSI B30.5 REQUIREMENTS

BOOM NO. 22C WITH OPEN THROAT TOP
146,400 LB. CRANE COUNTERWEIGHT
60,000 LB. CARBODY COUNTERWEIGHT
26'6" CRAWLERS EXTENDED

WARNING: This chart will apply only when two 12,000 lb. side ctwts. and two 30,000 lb. carbody ctwts. bear MEC registered Serial Numbers.

LIFTING CAPACITIES: Capacities for various boom lengths and operating radii may be based on percent of tipping, strength of structural components, operating speeds and other factors.

Capacities are for freely suspended loads and do not exceed 75% of a static tipping load. Capacities based on structural competence are shown by shaded areas.

Capacities are shown in pounds. Deduct 1200 pounds from capacities listed when single sheave upper boom point is attached and 1500 pounds when two sheave upper boom point is attached. To comply with B30.5 requirements, upper boompoint cannot be used on the 260 ft. boom. Weight of jib, (see chart A), all load blocks, hooks, weight ball, slings, hoist lines beneath boom and jib point sheaves, etc., is considered part of the main boom load. Boom is not to be lowered beyond radii where combined weights are greater than rated capacity. Where no capacity is shown, operation is not intended or approved.

OPERATING CONDITIONS: Machine to operate in a level position on a firm surface with crawlers fully extended and gantry in working position and be rigged in accordance with and under conditions referred to in rigging drawing No. 190693 and load line specification chard No. 6592-A.

Crane operator judgement must be used to allow for dynamic load effects of swinging, hoisting or lowering, travel, as well as adverse operating conditions & physical machine depreciation.

OPERATOR RADIUS: Operating is the horizontal distance form the axis of rotation to the center of vertical hoist line or load block with the load freely suspended. Add 14" to boom point radius for radius of sheave when using single part hoist line.

Boom angle is the angle between horizontal and centerline of boom butt and inserts and is an indication of operating radius. In all cases, operating radius shall govern capacity.

BOOM POINT ELEVATION: Boom point elevation, in feet, is the vertical distance from ground level to centerline of boom point shaft.

MACHINE EQUIPMENT: Machine equipped with 26'6" extendible crawlers, 48" treads, 17' retractable gantry, 12 part boom hoist reeving, four 1 3/8" boom pendants, 1st ctwt. 41,900 lbs., 2nd ctwt. 41,500 lbs., 3rd ctwt. 39,000 lbs., two 12,000 lbs. side ctwt's. and two 30,000 lbs. carbody ctwt's.

LOAD AND WHIP LINE SPECIFICATIONS

LOAD LINE:	1-1/8" - 6 x 31 Warrington-Seale, Extra Improved Plow Steel, Regular Lay, IWRC. Minimum Breaking Strength 65 Ton. (Approx. Weight Per Ft. in Lbs. 2.34)
WHIP LINE:	1-1/8" - Warrington-Seale, Improved Plow Steel, Regular Lay, IWRC. Minimum Breaking Strength 56.5 Ton. Maximum Load - 28,300 Lbs. Per Line. (Approx. Weight Per Ft. in Lbs. 2.34)

MAXIMUM BOOM AND JIB LENGTHS LIFTED UNASSISTED				DEDUCT FROM CAPACITIES WHEN JIB IS ATTACHED	
OVER FRONT OF BLOCKED CRAWLERS		OVER SIDE OF EXTENDED CRAWLERS			
BOOM LENGTH	JIB NO.123	BOOM LENGTH	JIB NO.123	JIB LENGTH	JIB NO.123
260'	--	260'	--	30'	3,000 lbs.
250'	--	250'	--	40'	3,600 lbs.
240'	40'	240'	40'	50'	4,200 lbs.
230'	60'	230'	60'	60'	4,900 lbs.
Load Block, hook and weight ball on ground at start.					

HOIST REEVING FOR MAIN LOAD BLOCK

No. Parts Of Line	1	2	3	4	5	6
Max Load - Lbs.	32,500	65,000	97,500	130,000	162,500	195,000
No. Parts of Line	7	8	9	10	11	12
Max. Load - Lbs.	227,500	260,000	292,500	325,000	357,500	400,000
No. Parts of Line	13					
Max. Load - Lbs.	430,000					

FOR JIB CAPACITIES, CONSULT JIB CHART.

BOOM LGTH FEET	OPER RAD. FEET	BOOM ANG DEG	BOOM POINT ELEV	CAPACITY: CRAWLERS EXTENDED
70	16.5	79.7	75.9	460,000
	17	79.3	75.8	400,000
	18	78.5	75.6	380,100
	19	77.6	75.4	363,000
	20	76.8	75.1	347,300
	22	75.1	74.6	319,600
	24	73.4	74.1	293,400
	26	71.7	73.5	266,100
	28	69.9	72.8	237,500
	30	68.2	72.0	214,300
	32	66.4	71.2	195,100
	34	64.6	70.2	178,900
	36	62.8	69.3	165,200
	38	60.9	68.2	153,300
	40	59.1	67.0	143,000
	45	54.1	63.7	122,100
	50	48.9	59.8	106,300
	55	43.2	54.9	93,900
	60	36.9	49.0	84,000
	65	29.4	41.3	75,800
	70	19.5	30.3	63,900

BOOM LGTH FEET	OPER RAD. FEET	BOOM ANG DEG.	BOOM POINT ELEV.	CAPACITY: CRAWLERS EXTENDED
80	17	80.6	85.9	392,800
	18	79.9	85.8	378,900
	19	79.2	85.6	361,800
	20	78.5	85.4	346,100
	22	77.0	84.9	318,400
	24	75.5	84.5	292,500
	26	74.0	83.9	265,600
	28	72.5	83.3	237,000
	30	71.0	82.7	213,800
	32	69.5	81.9	194,600
	34	68.0	81.2	178,500
	36	66.4	80.3	164,700
	38	64.8	79.4	152,800
	40	63.3	78.4	142,400
	45	59.2	75.7	121,500
	50	54.9	72.5	105,700
	55	50.4	68.6	93,300
	60	45.6	64.1	83,400
	65	40.3	58.8	75,200
	70	34.4	52.2	68,300
	75	27.4	43.9	52,500
	80	18.2	32.0	53,900

BOOM LGTH FEET	OPER RAD. FEET	BOOM ANG DEG.	BOOM POINT ELEV	CAPACITY: CRAWLERS EXTENDED
90	18	81.1	95.9	355,400
	19	80.4	95.7	346,900
	20	79.8	95.6	336,900
	22	78.5	95.2	317,400
	24	77.2	94.7	291,700
	26	75.9	94.3	264,800
	28	74.5	93.7	236,600
	30	73.2	93.2	213,400
	32	71.9	92.5	194,200
	34	70.5	91.9	178,000
	36	69.2	91.1	164,200
	38	67.8	90.3	152,300
	40	66.4	89.5	142,000
	45	62.9	87.1	121,100
	50	59.3	84.4	105,200
	55	55.5	81.2	92,800
	60	51.5	77.5	82,900
	65	47.3	73.2	74,700
	70	42.8	68.2	67,800
	75	37.9	62.3	62,000
	80	32.4	55.2	57,000
	85	25.8	46.2	52,600
	90	17.1	33.5	45,900

BOOM LGTH FEET	OPER. RAD. FEET	BOOM ANG DEG	BOOM POINT ELEV.	CAPACITY: CRAWLERS EXTENDED
100	19	81.4	105.9	332,900
	20	80.8	105.7	327,100
	22	79.6	105.4	316,200
	24	78.5	105.0	290,800
	26	77.3	104.5	263,900
	28	76.1	104.1	236,200
	30	74.9	103.6	212,900
	32	73.7	103.0	193,700
	34	72.5	102.4	177,500
	36	71.3	101.7	163,700
	38	70.1	101.0	151,800
	40	68.9	100.3	141,400
	45	65.8	98.2	120,500
	50	62.6	95.8	104,700
	55	59.3	93.0	92,300
	60	55.9	89.8	82,300
	65	52.4	86.2	74,100
	70	48.7	82.1	67,200
	75	44.8	77.4	61,400
	80	40.5	72.0	56,400
	85	35.9	65.6	52,000
	90	30.7	58.0	48,200
	95	24.5	48.5	44,900
	100	16.3	35.0	39,300

CAUTION! CHECK AMOUNT OF COUNTERWEIGHT ON MACHINE BEFORE USE OF THIS CHART.

 These load charts are intended for instructional purposes only. They were derived from manufacturer sales information which may not be complete or machine specific. Not responsible for typographical errors.

204F01.EPS

Figure 1 ◆ Load/capacity chart one.

LIFTCRANE CAPACITIES – 4100W SERIES 2 BOOM NO 22C WITH OPEN-THROAT TOP, CONTINUED

BOOM LGTH FEET	OPER. RAD. FEET	BOOM ANG. DEG.	BOOM POINT ELEV.	CAPACITY: CRAWLERS EXTENDED
110	22	82.3	115.7	240,500
	24	81.3	115.4	232,100
	26	80.2	115.0	224,400
	28	79.1	114.6	217,300
	30	78.1	114.1	210,700
	32	77.0	113.6	193,300
	34	75.9	113.1	177,100
	36	74.8	112.5	163,300
	38	73.7	111.8	151,400
	40	72.6	111.2	141,000
	45	69.9	109.3	120,100
	50	67.0	107.2	104,200
	55	64.1	104.7	91,800
	60	61.2	101.9	81,800
	65	58.1	98.8	73,600
	70	54.9	95.3	66,800
	75	51.6	91.3	60,900
	80	48.1	86.8	55,900
	85	44.4	81.7	51,600
	90	40.4	75.9	47,800
	95	36.1	69.2	44,400
	100	31.1	61.1	41,400
	105	25.3	51.1	38,700
	110	17.6	37.2	33,900
120	22	83.0	125.9	228,700
	24	82.0	125.5	220,500
	26	81.0	125.2	213,000
	28	80.0	124.8	206,100
	30	79.1	124.4	199,700
	32	78.1	123.9	192,800
	34	77.1	123.4	176,600
	36	76.1	122.9	162,800
	38	75.1	122.3	150,900
	40	74.1	121.7	140,500
	45	71.6	120.0	119,500
	50	69.1	118.1	103,700
	55	66.5	115.9	91,300
	60	63.8	113.4	81,300
	65	61.1	110.6	73,100
	70	58.3	107.5	66,200
	75	55.4	104.0	60,400
	80	52.4	100.1	55,300
	85	49.2	95.8	51,000
	90	45.9	91.0	47,200
	95	42.4	85.5	43,800
	100	38.6	79.4	40,800
	105	34.4	72.2	38,100
	110	29.7	63.7	35,700
	115	24.1	53.2	33,500
130	24	82.6	135.7	212,700
	26	81.7	135.3	205,300
	28	80.8	135.0	198,600
	30	79.9	134.6	192,300
	32	79.0	134.2	186,500
	34	78.1	133.7	176,300
	36	77.2	133.2	162,500
	38	76.3	132.7	150,600
	40	75.4	132.1	140,200
	45	73.1	130.6	119,300
	50	70.8	128.8	103,400
	55	68.4	126.8	91,000
	60	66.0	124.6	81,000
	65	63.5	122.0	72,800
	70	61.0	119.2	65,900
	75	58.4	116.1	60,000
	80	55.7	112.7	55,000
	85	53.0	108.9	50,700
	90	50.1	104.8	46,800
	95	47.1	100.1	43,500
	100	43.9	95.0	40,500
	105	40.6	89.2	37,800
	110	36.9	82.6	35,300
	115	33.0	75.1	33,100
	120	28.5	66.1	31,100
	125	23.1	55.1	29,300
140	26	82.3	145.5	198,400
	28	81.5	145.1	191,800
	30	80.7	144.8	185,600
	32	79.8	144.4	179,900
	34	79.0	144.0	174,700
	36	78.1	143.5	162,000
	38	77.3	143.0	150,100
	40	76.5	142.5	139,700
	45	74.3	141.1	118,700
	50	72.2	139.4	102,800
	55	70.0	137.6	90,400
	60	67.8	135.5	80,400
	65	65.6	133.2	72,200
	70	63.3	130.7	65,300
	75	60.9	127.9	59,500
	80	58.5	124.8	54,400
	85	56.1	121.4	50,100
	90	53.5	117.7	46,200
	95	50.9	113.7	42,900
	100	48.1	109.2	39,900
140	105	45.3	104.2	37,200
	110	42.2	98.8	34,700
	115	39.0	92.7	32,500
	120	35.5	85.7	30,500
	125	31.7	77.8	28,700
	130	27.4	68.5	27,000
	135	22.2	57.0	25,500
150	28	82.1	155.3	185,600
	30	81.3	154.9	179,600
	32	80.5	154.6	174,000
	34	79.7	154.2	168,800
	36	79.0	153.8	161,600
	38	78.2	153.3	149,600
	40	77.4	152.8	139,200
	45	75.4	151.5	118,200
	50	73.4	150.0	102,400
	55	71.4	148.3	89,900
	60	69.4	146.4	79,900
	65	67.3	144.3	71,700
	70	65.2	141.9	64,800
	75	63.0	139.4	59,000
	80	60.9	136.6	53,900
	85	58.6	133.5	49,600
	90	56.3	130.1	45,700
	95	54.0	126.5	42,400
	100	51.5	122.5	39,400
	105	49.0	118.2	36,700
	110	46.4	113.4	34,200
	115	43.6	108.2	32,000
	120	40.7	102.4	30,000
	125	37.6	96.0	28,200
	130	34.2	88.8	26,500
	135	30.5	80.5	25,000
	140	26.4	70.8	23,500
	145	21.4	58.8	22,200
160	28	82.6	165.4	179,800
	30	81.8	165.1	173,900
	32	81.1	164.7	168,400
	34	80.4	164.4	163,300
	36	79.7	164.0	158,600
	38	78.9	163.6	149,100
	40	78.2	163.1	138,700
	45	76.3	161.9	117,700
	50	74.5	160.5	101,800
	55	72.6	158.9	89,300
	60	70.7	157.1	79,300
	65	68.8	155.1	71,100
	70	66.8	153.0	64,200
	75	64.9	150.6	58,400
	80	62.9	148.0	53,300
	85	60.8	145.2	49,000
	90	58.7	142.2	45,100
	95	56.6	138.8	41,800
	100	54.4	135.2	38,700
	105	52.1	131.4	36,000
	110	49.7	127.1	33,600
	115	47.3	122.5	31,400
	120	44.8	117.5	29,400
	125	42.1	112.0	27,600
	130	39.3	105.9	25,900
	135	36.3	99.2	24,300
	140	33.1	91.7	22,900
	145	29.5	83.0	21,600
	150	25.5	72.9	20,400
	155	20.7	60.5	19,200
170	30	82.3	175.2	169,000
	32	81.6	174.9	163,600
	34	81.0	174.5	158,700
	36	80.3	174.2	154,000
	38	79.6	173.8	148,800
	40	78.9	173.3	138,400
	45	77.2	172.2	117,400
	50	75.4	170.9	101,500
	55	73.7	169.4	89,100
	60	71.9	167.7	79,000
	65	70.1	165.9	70,800
	70	68.3	163.9	63,900
	75	66.4	161.7	58,100
	80	64.6	159.3	53,000
	85	62.7	156.7	48,700
	90	60.8	153.9	44,900
	95	58.8	150.8	41,500
	100	56.8	147.5	38,500
	105	54.7	144.0	35,800
	110	52.6	140.2	33,300
	115	50.4	136.0	31,100
	120	48.1	131.5	29,100
	125	45.8	126.7	27,300
	130	43.3	121.4	25,600
	135	40.8	115.7	24,100
	140	38.1	109.3	22,600
	145	35.2	102.3	21,300
	150	32.0	94.5	20,100
	155	28.6	85.5	18,900
	160	24.7	75.1	17,800
	165	20.0	62.2	16,400
180	32	82.1	185.0	158,700
	34	81.5	184.7	154,000
	36	80.8	184.3	149,500
	38	80.2	184.0	145,200
	40	79.5	183.6	137,900
	45	77.9	182.5	116,900
	50	76.3	181.2	101,000
	55	74.6	179.8	88,500
	60	72.9	178.3	78,500
	65	71.3	176.6	70,300
	70	69.6	174.7	63,400
	75	67.8	172.6	57,500
	80	66.1	170.4	52,500
	85	64.3	168.0	48,100
	90	62.5	165.4	44,300
	95	60.7	162.5	40,900
	100	58.9	159.5	37,900
	105	57.0	156.2	35,200
	110	55.0	152.7	32,700
	115	53.0	149.0	30,500
	120	51.0	144.9	28,500
	125	48.9	140.5	26,700
	130	46.7	135.8	25,000
	135	44.4	130.7	23,500
	140	42.0	125.2	22,000
	145	39.6	119.2	20,700
	150	36.9	112.6	19,500
	155	34.1	105.3	18,300
	160	31.1	97.2	17,200
	165	27.7	87.9	16,200
	170	23.9	77.1	15,300
	175	19.4	63.8	13,500
190	32	82.5	195.1	153,100
	34	81.9	194.8	148,800
	36	81.3	194.5	144,800
	38	80.7	194.1	140,800
	40	80.1	193.8	137,000
	45	78.5	192.7	116,400
	50	77.0	191.5	100,500
	55	75.4	190.2	88,000
	60	73.9	188.8	78,000
	65	72.3	187.1	69,700
	70	70.7	185.4	62,800
	75	69.1	183.4	57,000
	80	67.4	181.4	52,000
	85	65.8	179.1	47,600
	90	64.1	176.6	43,800
	95	62.4	174.0	40,400
	100	60.7	171.2	37,400
	105	58.9	168.2	34,700
	110	57.1	164.9	32,200
	115	55.3	161.4	30,000
	120	53.4	157.7	28,000
	125	51.5	153.7	26,200
	130	49.5	149.5	24,500
	135	47.5	144.9	22,900
	140	45.3	140.0	21,500
	145	43.1	134.6	20,200
	150	40.8	128.9	18,900
	155	38.4	122.6	17,800
	160	35.9	115.8	16,700
	165	33.2	108.3	15,700
	170	30.2	99.8	14,800
	175	27.0	90.3	13,900
	180	23.3	79.1	13,100
	185	18.9	65.4	11,000
200	34	82.3	204.9	143,100
	36	81.7	204.6	139,500
	38	81.2	204.3	136,000
	40	80.6	203.9	132,500
	45	79.1	203.0	115,900
	50	77.7	201.8	100,000
	55	76.2	200.6	87,500
	60	74.7	199.2	77,400
	65	73.2	197.7	69,200
	70	71.7	196.0	62,300
	75	70.2	194.2	56,400
	80	68.6	192.2	51,400
	85	67.1	190.1	47,000
	90	65.5	187.8	43,200
	95	63.9	185.3	39,800
	100	62.3	182.7	36,800
	105	60.6	179.8	34,100
	110	59.0	176.8	31,600
	115	57.3	173.6	29,400
	120	55.5	170.2	27,400
	125	53.0	166.5	25,600
	130	51.9	162.6	23,900
	135	50.1	158.4	22,300
	140	48.1	153.9	20,900
	145	46.2	149.1	19,600
	150	44.1	144.0	18,300
	155	42.0	138.4	17,200
	160	39.7	132.5	16,100
	165	37.4	126.0	15,100
	170	34.9	118.9	14,200
200	175	32.3	111.1	13,300
	180	29.4	102.4	12,400
	185	26.2	92.5	11,700
	190	22.6	81.0	10,900
	195	18.4	67.0	8,500
210	36	82.1	214.7	134,200
	38	81.6	214.4	131,000
	40	81.0	214.1	127,900
	45	79.6	213.2	115,500
	50	78.3	212.1	99,600
	55	76.9	210.9	87,100
	60	75.4	209.6	77,100
	65	74.0	208.1	68,800
	70	72.6	206.6	61,900
	75	71.2	204.8	56,100
	80	69.7	203.0	51,000
	85	68.2	201.0	46,700
	90	66.7	198.8	42,800
	95	65.2	196.5	39,400
	100	63.7	194.0	36,400
	105	62.2	191.3	33,700
	110	60.6	188.5	31,300
	115	59.0	185.5	29,100
	120	57.4	182.3	27,100
	125	55.7	178.9	25,200
	130	54.1	175.2	23,500
	135	52.3	171.4	22,000
	140	50.6	167.3	20,600
	145	48.8	162.9	19,200
	150	46.9	158.2	18,000
	155	45.0	153.2	16,800
	160	43.0	147.9	15,800
	165	40.9	142.1	14,800
	170	38.7	135.9	13,800
	175	36.5	129.2	12,900
	180	34.0	121.9	12,100
	185	31.5	113.9	11,300
	190	28.7	104.9	10,600
	195	25.6	94.7	9,900
	200	22.1	82.9	8,800
	205	17.9	68.5	6,500
220	38	82.0	224.6	125,700
	40	81.4	224.2	122,800
	45	80.1	223.3	115,000
	50	78.8	222.3	99,100
	55	77.5	221.2	86,600
	60	76.1	220.0	76,600
	65	74.8	218.6	68,300
	70	73.4	217.1	61,400
	75	72.0	215.4	55,500
	80	70.7	213.7	50,500
	85	69.3	211.8	46,100
	90	67.9	209.7	42,300
	95	66.4	207.5	38,900
	100	65.0	205.2	35,800
	105	63.6	202.7	33,100
	110	62.1	200.0	30,700
	115	60.6	197.2	28,500
	120	59.1	194.2	26,500
	125	57.5	191.0	24,600
	130	55.9	187.6	23,000
	135	54.3	184.0	21,400
	140	52.7	180.2	20,000
	145	51.0	176.1	18,600
	150	49.3	171.8	17,400
	155	47.6	167.3	16,300
	160	45.7	162.4	15,200
	165	43.9	157.2	14,200
	170	41.9	151.7	13,200
	175	39.9	145.7	12,300
	180	37.8	139.3	11,500
	185	35.6	132.4	10,700
	190	33.2	124.8	10,000
	195	30.7	116.6	9,300
	200	28.0	107.3	8,600
	205	25.0	96.9	8,000
	210	21.5	84.8	6,500

CAPACITIES CONTINUED ON NEXT PAGE

These load charts are intended for instructional purposes only. They were derived from manufacturer sales information which may not be complete or machine specific. Not responsible for typographical errors.

204F02.EPS

Figure 2 ◆ Load/capacity chart two.

WEIGHTS 4100W Cont'd

DESCRIPTION	APPROX. WEIGHT (IN LBS.)
JIB NO. 123	
Jib Top - 15' (w/Jib point)	695
Jib Butt - 15'	690
Jib Insert - 10'	340
Basic Pendant - 33' 3-3/4" (2 Req'd)	115 each
Pendant - 10' (2 Per insert)	65 each
Jib Backstay Pendant	155 each
Jib Strut - 12'- 6"	365
COMPONENTS	
Hook Rollers (6) - w/shafts	1,020
Light Plant - 6.5KW - w/mounting platform	1,390
Catwalk - left and right side w/rails	1,320
Lagging - 27-5/8" dia. plain	1,410
Boom Hoist Rope - 12 Part - 760' of 7/8" - 6 x 26	1,080
Wire Rope Guide Assembly - Lower	325
Wire Rope Guide Assembly - Upper	510
Rope Guide Roller Assembly	55 each
2-Part Gantry w/Telescopic Backhitch	7,805
Equalizer	2,000
Hoist Line - 1-1/8" - 6 x 31	2.34 lbs./ft.
Whip Line - 1-1/8" - 6 x 31	2.34 lbs./ft.
15-Ton Hook and Weight Ball	865
100-Ton Hook Block Assembly	2,065
200-Ton Hook Block Assembly	4,900
230-Ton Hook Block Assembly	5,375
Boom Stop - Telescopic Air Cushioned	675
Dragline Fairlead - Revolving	1,910
Dragline Fairlead - Hinged	9,330

NOTE: The above weights may fluctuate up or down 5% due to manufacturing tolerances.

204F03.EPS

Figure 3 ◆ Load/capacity chart three.

3.2.0 Multiple-Crane Lifting

Multiple-crane lifts are always considered critical lifts, requiring detailed plans and procedures. The effects of the relative motion between cranes may induce additional weight on the cranes, the load, and the lifting gear. Multiple-crane lifting must be used when the physical dimensions, characteristics, mass, or required movement of the load prevent the operation from being carried out by a single crane. Procedures must be in the written form of a lift plan and signed by qualified personnel. These factors should be considered when planning multiple-crane lifts:

- The lift must be planned and carried out under qualified supervision.
- Ground conditions must be stable and level.
- Consideration must be given to the environment and the location of the lift.
- All cranes must rest on a solid, firm, level surface.

- All cranes must be level.
- The load weight must be known.
- The longest load radius for each crane of the operation must be measured.
- The maximum allowable boom length operating radius and minimum boom angle must be known.
- The rated capacity of each crane must be known.
- Because loads transfer rapidly, no crane should be loaded to more than 75 percent of its net capacity.
- Exactly how much of the load to be carried by each crane must be known. The rigging must be arranged to divide the load as planned.
- The line, swing, and boom speeds of the cranes must be matched.
- Swing and booming methods should be kept to a minimum.
- Whenever possible, the cranes should not travel with the load.
- A lift plan must be developed.
- Signal persons, riggers, and operators must know exactly what to do before the lift begins.
- If possible, a dry run should be conducted without a load.
- Communications by radio are preferred when hand signals are not effective.
- One person is required to direct and control the operation.
- Hoist lines must be kept vertical at all times. When the lines are not vertical, the cranes are transferring load to each other, and one crane may sideload the other.

WARNING!

Multiple-crane lifts are particularly hazardous. Because the movements of multiple cranes must be closely coordinated, all crane movements must be carefully planned and slowly executed to prevent lift failures. All personnel involved must be thoroughly familiar with the lift plan and any applicable emergency procedures.

Multiple cranes are often used to lift very heavy loads. *Figure 4* shows a three-crane lift with two of the cranes using an equalizer beam with a special load attachment. The equalizer beam and attachment split the load between the two cranes. Equalizer beams are covered later in this module.

Whenever two or more cranes are used to lift a load, the location of each crane's hook block above

204F04.EPS

Figure 4 ◆ Three-crane lift.

the load has to be chosen carefully so that the load is shared between the cranes. Because each crane used will usually have a different net capacity, the method for properly locating each crane's hook block above the load (*Figure 5*) for a two-crane lift is as follows:

Step 1 Determine the lowest value of rated net capacity that Crane One will have during the whole operation. When lifting from the main boom or an extension, consider all factors, including configuration, quadrant of operation, boom length, boom angle, load radius, weight of rigging, and all capacity deductions.

Step 2 Repeat Step 1 for Crane Two.

Step 3 Ensure that the sum of the lowest rated net capacities of the cranes exceeds the load weight.

Step 4 Locate the position of the load's center of gravity.

Step 5 Each crane's load blocks must be positioned relative to the load's center of gravity according to the following formula:

$$\text{(Net capacity of Crane One)} \times A = \text{(net capacity of Crane Two)} \times B$$

Step 6 The crane with the greatest net capacity must always be closest to the load's center of gravity. Measure the distance of this point from the center of gravity. If you have measured distance A, then:

$$B = \frac{\text{net capacity of Crane One}}{\text{net capacity of Crane Two}} \times A$$

If you have measured distance B then:

$$A = \frac{\text{net capacity of Crane Two}}{\text{net capacity of Crane One}} \times B$$

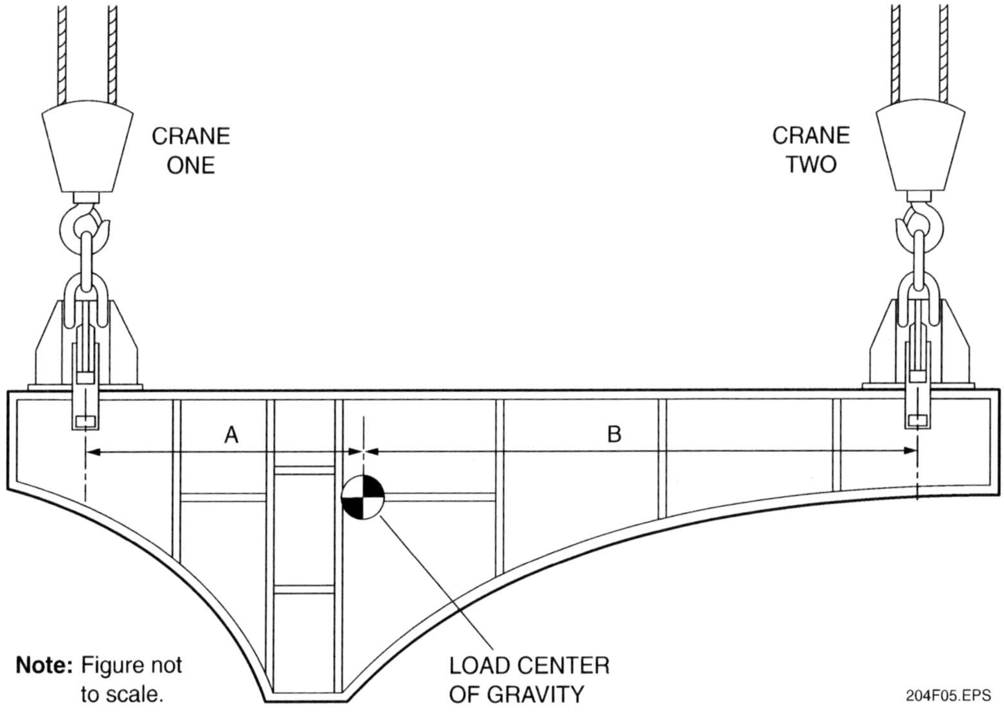

Figure 5 ◆ Multiple-crane lift.

Step 7 Determine the share of the load weight that will be carried by each crane.

Load on Crane One

$$\frac{B}{A + B} \times \text{load weight}$$

Load on Crane Two

$$\frac{A}{A + B} \times \text{load weight}$$

Step 8 Ensure that neither crane is loaded beyond its lowest rated net capacity.

NOTE

Each lifting operation is a unique situation. The numbers provided in this example are randomly selected for instructional purposes.

An example of this procedure is as follows:

Step 1 Lowest rated net capacity of Crane One = 156,000 lbs – 500 lbs (beam clamp) = 155,500 lbs.

Step 2 Lowest rated net capacity of Crane Two = 90,500 lbs – 500 lbs (beam clamp) = 90,000 lbs.

Step 3 Net capacity = 155,500 lbs + 90,000 lbs = 245,500 lbs.

Step 4 Locate the load's center of gravity from the ends of the load, as shown in *Figure 6*.

Step 5 To set the location of the hook blocks, use the formula:

$$(\text{Net capacity of Crane One}) \times A = (\text{net capacity of Crane Two}) \times B$$

Step 6 Choose a hypothetical length for A that will be within the length of C in *Figure 6*. Start with A = 45 ft (*Figure 7*), for example.

$$\text{Calculate } B = \frac{\text{net capacity of Crane One}}{\text{net capacity of Crane Two}} \times A$$

$$= \frac{155{,}500 \text{ lbs}}{90{,}000 \text{ lbs}} \times 45 \text{ ft}$$

$$= 78 \text{ feet}$$

This configuration will not work because the hook block for B (78 ft) would be beyond the end of the right side of the load (*Figure 8*). Choose another dimension for A: 36 ft (*Figure 9*).

$$\text{Recalculate } B = \frac{\text{net cap. of Crane One}}{\text{net cap. of Crane Two}} \times A$$

$$= \frac{155{,}500 \text{ lbs}}{90{,}000 \text{ lbs}} \times 36 \text{ ft}$$

$$= 62.2 \text{ ft}$$

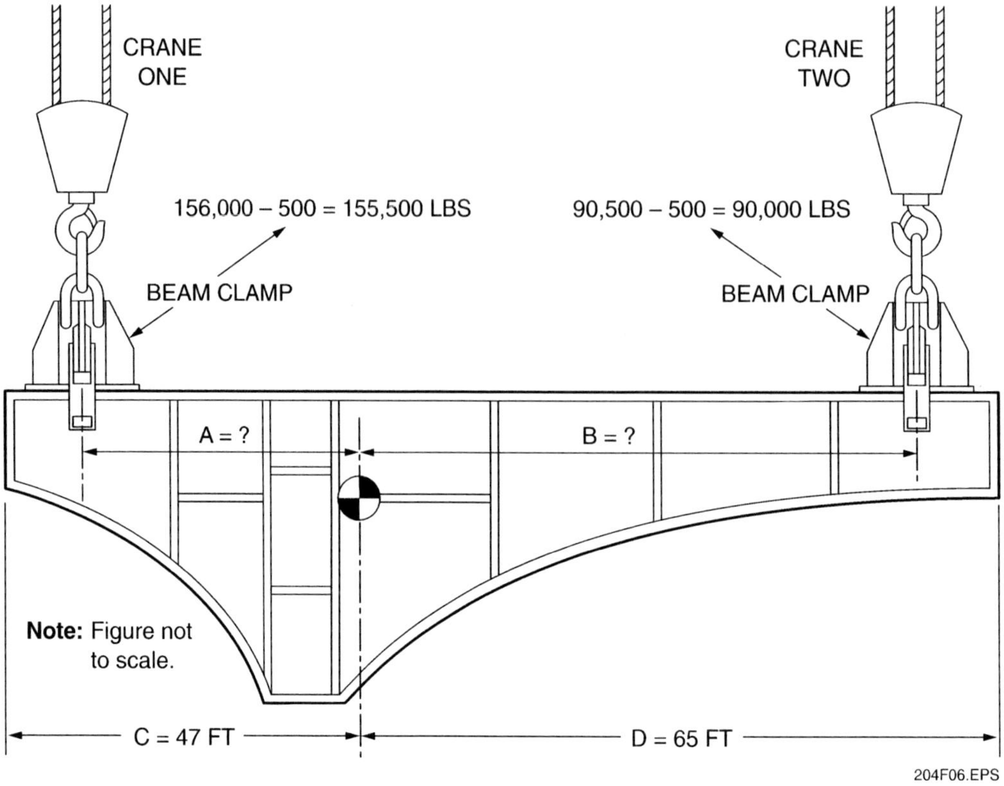

Figure 6 ◆ Multiple-crane lift – center of gravity location.

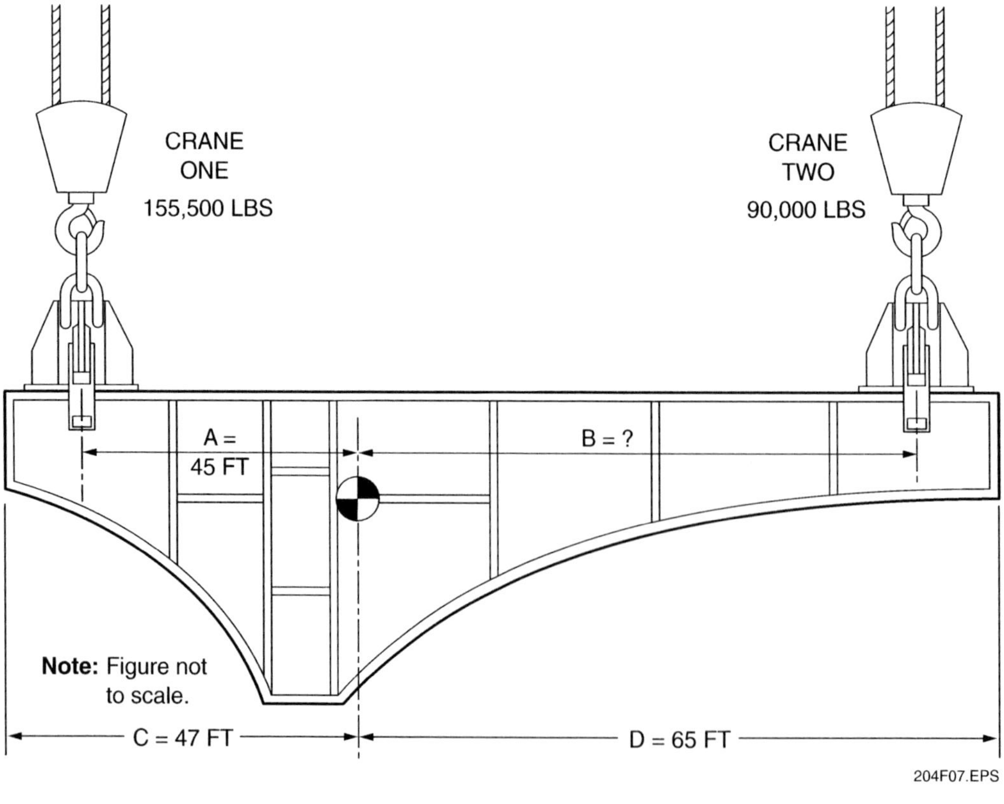

Figure 7 ◆ Multiple-crane lift – distance B, trial 1.

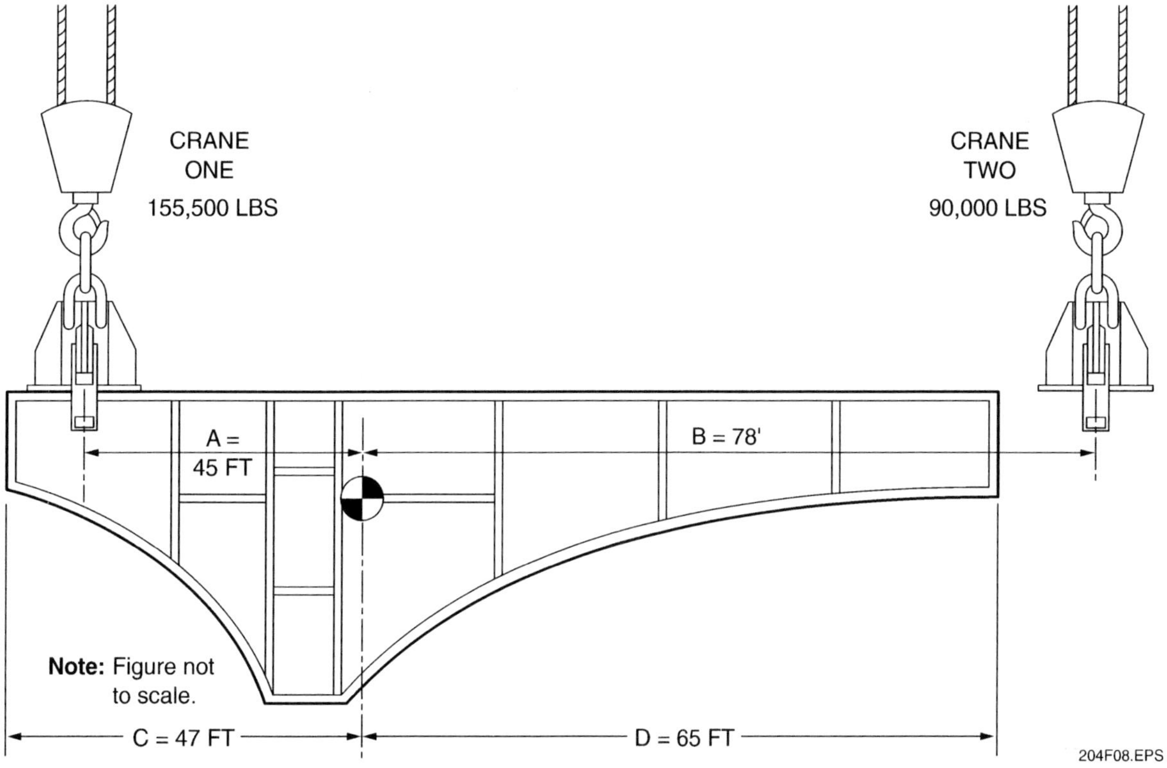

Figure 8 ◆ Multiple-crane lift – distance B, trial 1 results.

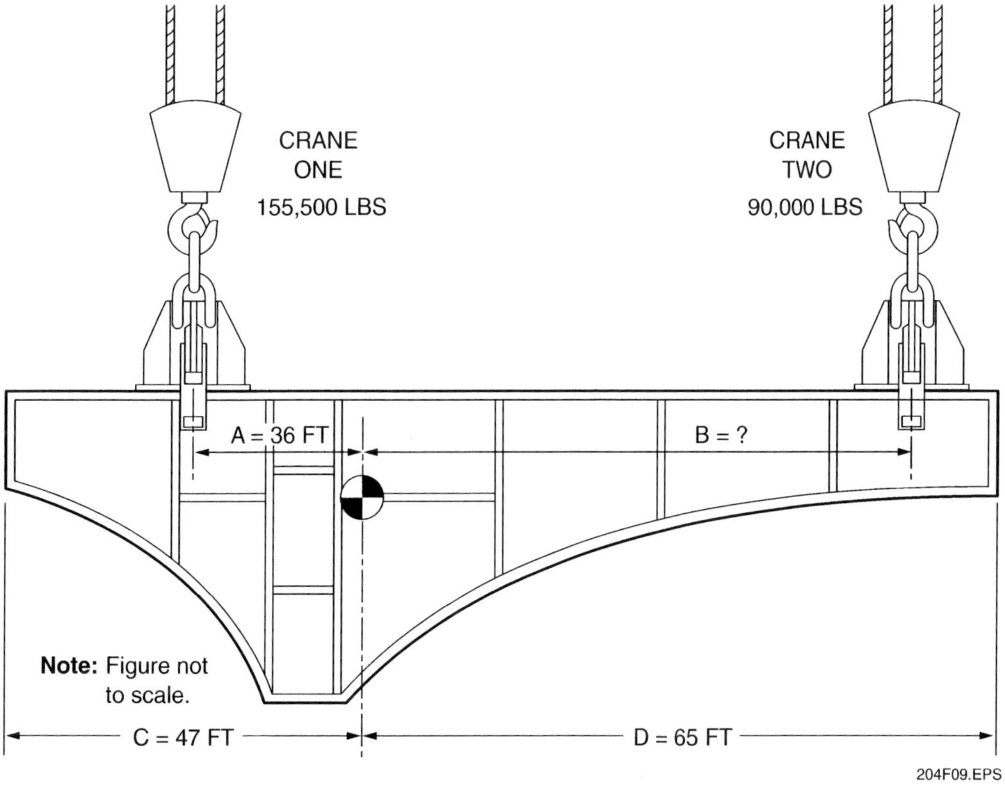

Figure 9 ◆ Multiple-crane lift – distance B, trial 2.

This dimension fits the load (*Figure 10*).

Step 7 Determine the actual load that each crane will be carrying:

$$\text{Load/Crane One} = \frac{B}{A + B} \times \text{load weight}$$

$$= \frac{62.2}{36 + 62.2} \times 180{,}000 \text{ lbs}$$

$$= 114{,}012.2 \text{ lbs}$$

This load is less than Crane One's rated net capacity of 155,500 lbs, and is under 75 percent (73.3 percent) of the capacity.

$$\text{Load/Crane Two} = \frac{A}{A + B} \times \text{load weight}$$

$$= \frac{36}{36 + 62.2} \times 180{,}000 \text{ lbs}$$

$$= 65{,}988 \text{ lbs}$$

This load is less than Crane Two's rated net capacity of 90,000 lbs and is under 75 percent (73.3 percent) of the capacity.

Step 8 Since neither crane is overloaded nor loaded beyond 75 percent of its capacity, the lift can proceed with A = 36 ft and B = 62.2 ft. Crane One will be 36 ft from the load's center of gravity, and Crane Two will be 62.2 ft from it.

NOTE

If the load radius of either crane is changed, repeat all of the steps using the new net capacity of the crane(s).

Multiple crane lifts can also use **equalizer** beams (*Figure 11*). Equalizer beams distribute the weight of a load between two cranes. By adjusting the crane and/or load attachment points on the beam, it is possible to share the load weight between the cranes in any proportion desired.

The operator can determine the location on the equalizer beam where the load is to be picked up, allowing each crane to share the load proportionately to its capacity, by following this method:

Step 1 Determine the lowest value of rated net capacity that Crane One will have when lifting from the main boom or an extension during the whole operation. Consider all factors, including configuration, operational quadrant, and load radius. Remember the load radius of each crane is measured to its pick-up point on the equalizer, not to the center of gravity of the load.

Step 2 Repeat Step 1 for Crane Two.

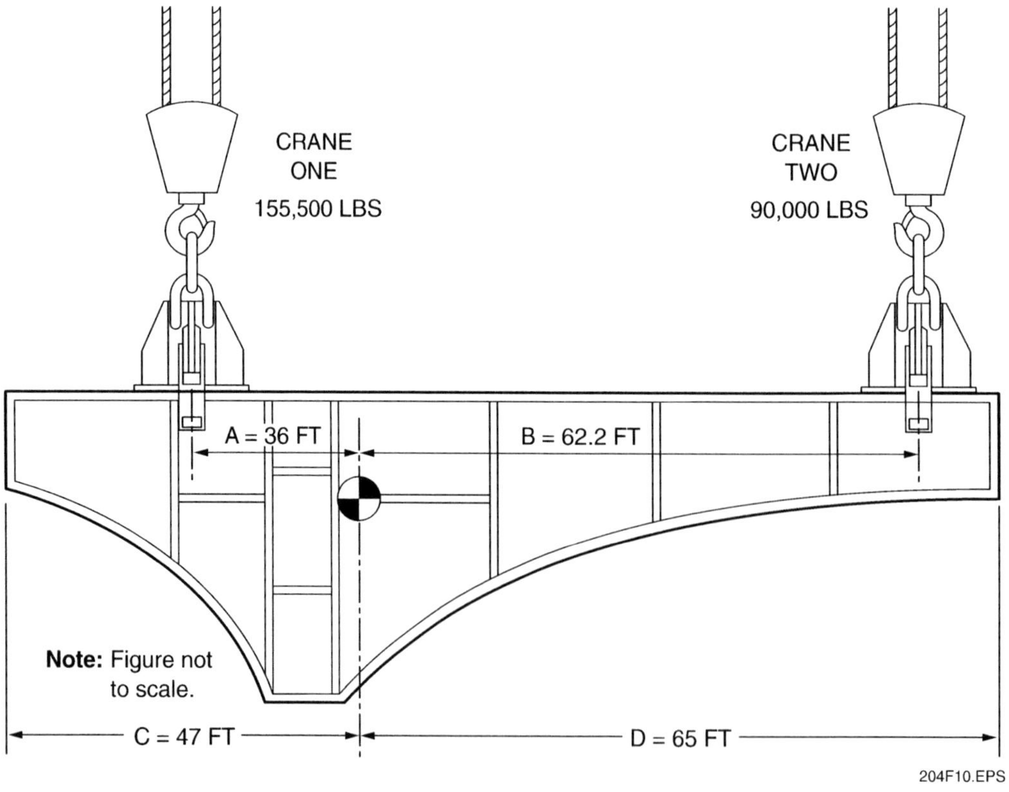

Figure 10 ◆ Multiple-crane lift – distance B, trial 2 results.

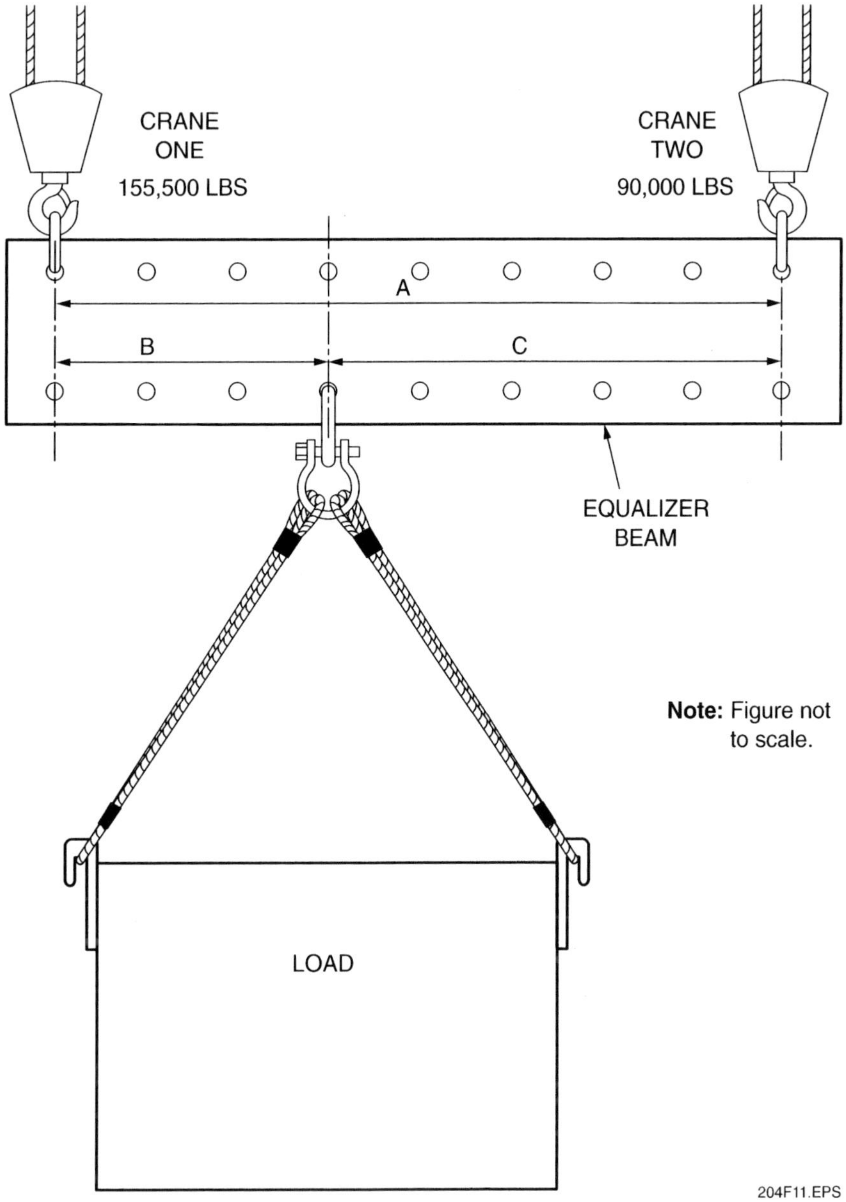

Figure 11 ◆ Multiple-crane lift with equalizer beam.

Step 3 Check to ensure that the sum of the lowest rated net capacities of the cranes exceeds the load weight plus the equalizer and sling weight. Ideally the weight should not exceed, 75 percent of the sum of their capacities.

Step 4 Measure the distance between the crane pick-up points on the equalizer beam (A).

Step 5 Establish the attachment point of the load on the equalizer beam according to the following formulas:

$$B = \frac{\text{net capacity of Crane Two}}{\text{total net capacity of Cranes One and Two}} \times A$$

$$C = \frac{\text{net capacity of Crane One}}{\text{total net capacity of Cranes One and Two}} \times A$$

Step 6 When the equalizer pick-up points have been established, determine the share of the load weight, including the weight of the equalizer, that will be carried by each crane:

$$\text{Load on Crane One} = \frac{C}{A} \times \text{(load weight)}$$

$$\text{Load on Crane Two} = \frac{B}{A} \times \text{(load weight)}$$

Check to ensure that neither crane is loaded beyond its lowest rated net capacity. Remember, the best situation is where neither crane is loaded in excess of 75 percent of its lowest rated net capacity. An example of this procedure follows (see *Figure 12*).

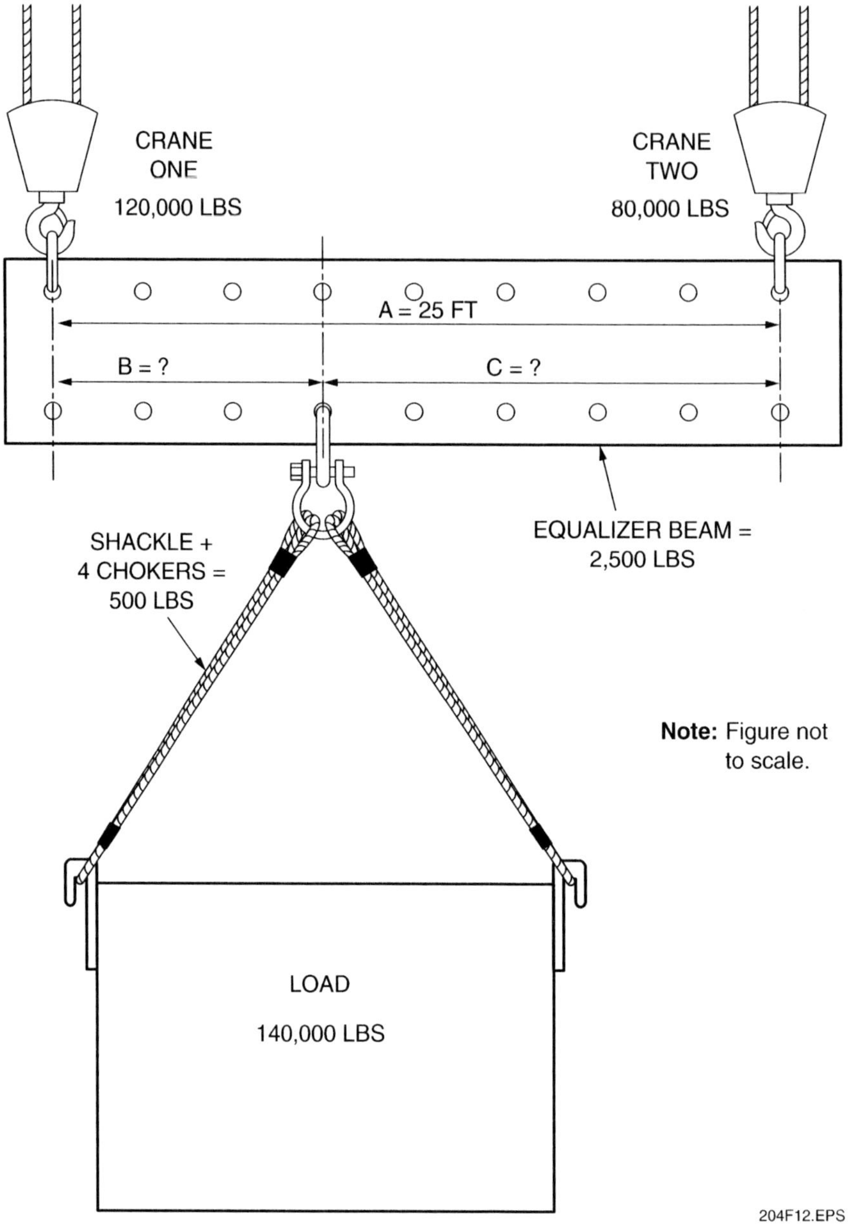

Figure 12 ◆ Multiple-crane lift with equalizer beam, determining distance B and C.

Step 1 Check the total capacity of the cranes.

Crane One = 120,000 lbs,
Crane Two = 80,000 lbs

Crane One + Crane Two = total capacity

120,000 lbs + 80,000 lbs = 200,000 lbs total capacity

This is greater than the total load weight of 143,000 lbs (load + equalizer). Also, 75 percent of the capacities' sum exceeds the load weight (75 percent of 200,000 = 150,000 lbs).

Step 2 Establish the load's pick-up point.

$$B = \frac{\text{net capacity of Crane Two}}{\text{total net capacity of Cranes One and Two}} \times A$$

$$B = \frac{80{,}000 \text{ lbs}}{80{,}000 \text{ lbs} + 120{,}000 \text{ lbs}} \times 25 \text{ ft}$$

$$B = \frac{80{,}000 \text{ lbs}}{200{,}000 \text{ lbs}} \times 25 \text{ ft}$$

B = 10 ft; therefore,

C = 15 ft, as shown in *Figure 13*

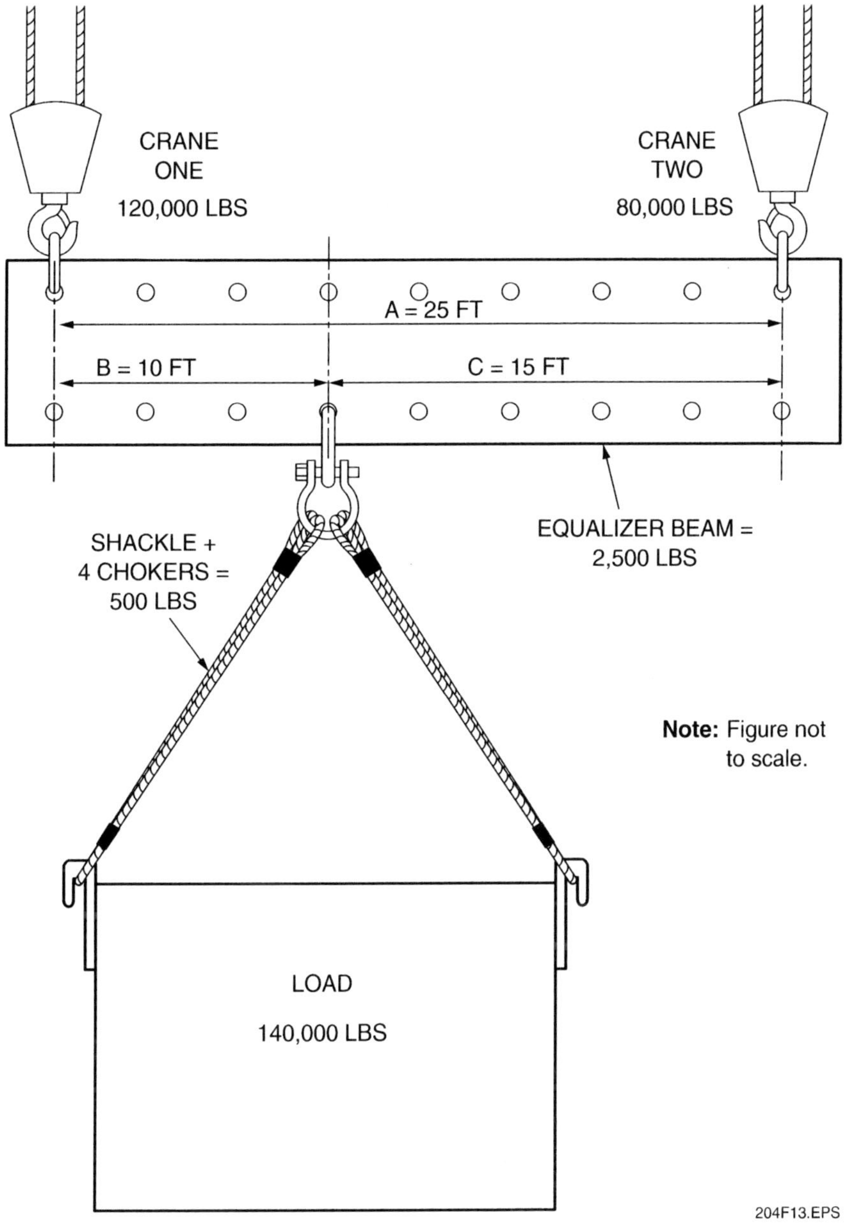

Figure 13 ◆ Multiple-crane lift with equalizer beam, distance B and C results.

Step 3 Determine the load that each crane will actually be carrying.

$$\text{Load on Crane One} = \frac{C}{A} \times (\text{total load weight})$$

$$= \frac{15}{25} \times 143{,}000 \text{ lbs}$$

$$= 85{,}800 \text{ lbs}$$

This number is less than the lowest rated net capacity and also meets the 75-percent capacity guideline (75 percent of 120,000 = 90,000 lbs).

$$\text{Load on Crane Two} = \frac{B}{A} \times (\text{total load weight})$$

$$= \frac{10}{25} \times 143{,}000 \text{ lbs}$$

$$= 57{,}200 \text{ lbs}$$

This number is less than the lowest rated net capacity and also meets the 75-percent capacity guideline (75 percent of 80,000 = 60,000 lbs).

4.0.0 ◆ CRITICAL LIFTS

A designated person must classify each lift into one of the following categories (ordinary, critical, or pre-engineered production) prior to planning the lift. A lift must be designated as a critical lift if collision, upset, or dropping could result in any one of the following:

- Unacceptable risk of personnel injury or significant adverse health impact (on site or off site)
- Significant release of radioactive or other hazardous material or other undesirable conditions
- Undetectable damage that would jeopardize future operations or the safety of a facility
- Damage that would result in unacceptable delay to schedule or other significant program impact, such as loss of vital data

A lift should also be designated as critical if the load requires exceptional care in handling because of size, weight, close-tolerance installation, high susceptibility to damage, or other unusual factors.

For critical lifts, the operating organization must appoint a rigging superintendent or other qualified person who must be present at the lift site during the entire lifting operation. The rigging supervisor or other qualified person must ensure that a lift plan or procedure (see the *Appendix*) is prepared that defines the operation and includes the following:

- Identification of the items to be moved, the weight, dimensions, and center of gravity of the load, and any hazardous or toxic materials that are present
- Identification of operating equipment to be used by type and rated capacity
- Rigging sketches that include (as applicable):
  - Identification and rated capacity of slings, lifting bars, rigging accessories, and below-the-hook lifting devices
  - Load-indicating devices
  - Lifting points
  - Sling angles
  - Boom and swing angles
  - Methods of attachment
  - Crane orientations
  - Other factors affecting equipment capacity
- Operating procedures and special instructions to operators, including rigging precautions and safety measures to be followed as applicable

Experienced operators who have been trained and qualified to operate the specific equipment to be used must be assigned to make the lift.

Only designated, qualified signalers must give signals to the operator. However, the operator must obey a STOP signal at all times, no matter who gives the signal.

The procedure and rigging sketches must be reviewed and approved by the responsible manager (or designee) and the responsible oversight organization (such as safety, quality assurance, or quality control) before the lift is made.

5.0.0 ◆ ENGINEERING CONSIDERATIONS

The following **engineering** requirements need to be considered when planning a lift:

- Basic load data, such as the object to be lifted, its dimensions and general configuration (rectangle, square, sphere, or irregular), and its gross weight
- Location in which load is to be lifted and placed
- Maximum operating radius
- Lift attachment locations and descriptions
- Maximum distance between any two attachment locations
- Sling lengths needed to keep the horizontal angle greater than 60 degrees
- Specifications of the crane to be used (manufacturer, model)
- Counterweight attachments

Lattice boom cranes:

- Counterweights used on the lower and upper works
- Number of boom sections used and their sizes and lengths
- Number of jib sections used and their sizes and lengths
- Maximum capacity at the maximum lift radius
- Hoist block capacity
- Wire rope size and weight per foot
- Rated single line pull
- Parts of line reeved to the block
- The maximum capacity of the lift point for the boom and the jib

Ground support:

- Ground pressure allowed at the setup location and the ground pressure created by the crane and load at all radii out to the planned maximum load radius
- The maximum allowable crane load that will not exceed the ground support

Load calculations:

- Weight of the load, crane components below the lifting points, and wire rope

Administration:

- Pre-lift meeting
- Completion of lift plan

6.0.0 ◆ LIFT PLAN IMPLEMENTATION

For a lift plan to be successful, the critical data must be communicated to all personnel involved. Holding a pre-lift meeting accomplishes this goal. The signal person(s), riggers, operators, management, and site supervisors should all be involved in this meeting. Officially identify which operator and signal person(s) will be assigned the job, and determine how communication is to be maintained during the lift (including the use of hand signals, radio, or both, or other methods). Also, determine whether tag lines will be used. Discuss all potential hazards, including weather, power lines, or mechanical failure. Review procedures in the event of an emergency. Finally, have the person responsible for the lift, the site superintendent, the site safety manager, and any others needed sign off on the lift plan.

If required by the critical lift procedure, a test lift must be done before the critical lift. Conditions for a test lift should closely simulate actual conditions, which may include weight, rigging selection and configuration, and other relevant factors.

Summary

Lift plans are just as important to the success of a lifting operation as proper load/capacity chart interpretation. The operator, supervisor, signal person, and all other personnel involved in the lift must understand what is required on a lift plan and how it affects them. Personnel must also understand mathematical calculations and the equipment's limitations.

Review Questions

1. A document that outlines all weight, rigging, and configuration requirements for a lift is a(n) _____.
   a. lift plan
   b. manufacturer's worksheet
   c. pre-lift qualification
   d. load chart

2. Lift plans are mandatory for _____.
   a. steel erection or multiple-crane lifts
   b. lifts using extensions or jibs
   c. luffing or derricking operations
   d. lifts using a load block

3. The type of hitch and number of shackles are criteria that fall under the category of _____.
   a. load weight
   b. sizing of slings
   c. wire ropes
   d. rope reeving

4. Foundation and location of electrical hazards are examples of _____ information.
   a. crane placement
   b. hoisting
   c. hydraulic
   d. pneumatic

5. The number of feet and the weight per foot of the _____ need to be considered as part of a lift plan.
   a. boom
   b. jib
   c. wire rope
   d. swing-away

6. Over-the-front, over-the-rear, and over-the-side are _____ considerations for crane placement.
   a. crawler configuration
   b. area of operation
   c. gantry position
   d. counterweight

7. Critical lifts always require _____.
   a. an equalizer beam
   b. beam clamps
   c. radio communication
   d. a lift plan

8. When lifting from the main boom of a lattice crane, an erected jib, whip line length below the jib, and the length of the line from the lift point to load block are calculations involving the _____.
   a. weight of crane components below the lifting point
   b. load block height
   c. rigging on the block
   d. effective weight of crane components past the lifting point

9. The load block weight is considered part of the weight of the _____.
   a. load to be lifted, not including rigging
   b. crane components below the lifting point
   c. crane components past the lifting point
   d. boom components

10. The effective weight of crane components past a main boom lifting point includes _____.
    a. an erected jib
    b. any parts of line considered part of the load
    c. the maximum wire rope weight to be experienced during the lift
    d. lifting beams, slings, and shackles

11. If the net capacity of a crane in a particular configuration is 95,000 lbs and the total weight of the deductions is 2,500 lbs, the gross capacity is _____ lbs.
    a. 92,500
    b. 95,000
    c. 97,500
    d. impossible to calculate without more information

12. If the gross capacity of a crane at 200 ft of boom length and 50 ft of load radius is 83,000 lbs, and the deductions are a 1,500-lb boom point, a 865-lb hook block, 335 lbs of rigging, and 1,900 lbs of rope, the net capacity is _____ lbs.
    a. 74,600
    b. 78,400
    c. 87,600
    d. impossible to calculate without more information

13. If the gross capacity of a crane is 90,000 lbs and its net capacity in a particular configuration is 83,425 lbs, the total weight of the deductions is _____ lbs.
    a. 2,875
    b. 4,932
    c. 6,575
    d. impossible to calculate without more information

14. By adjusting the crane and/or load attachment points on an equalizer beam, it is possible to _____.
    a. safely load a crane beyond its net capacity
    b. ensure that load rate exceeds the lowest rated net capacities of the cranes
    c. share the load between cranes in any proportion desired
    d. share the load between cranes equally

15. In a multiple-crane lift, the crane with the greatest net capacity must always be _____.
    a. equipped with single part hoist line
    b. loaded to at least 75 percent of its rated capacity
    c. farthest from the load's center of gravity
    d. closest to the load's center of gravity

16. When an equalizer beam is used, the load radius of each crane is measured to _____.
    a. the center of gravity of the load
    b. its pick-up point on the equalizer
    c. the center of the equalizer
    d. the attachment point of the load on the equalizer

17. In a multiple-crane lift using an equalizer, the weight of the equalizer is considered part of the load weight.
    a. True
    b. False

18. In a multiple-crane lift, if the total load weight is 100,000 lbs, and the total net capacity of Crane One is 80,000 lbs, the total net capacity of Crane Two must be no less than _____ lbs to meet the 75-percent capacity guidelines.
    a. 53,333
    b. 75,050
    c. 100,500
    d. 180,000

Refer to *Figure 1* for Questions 19 and 20.

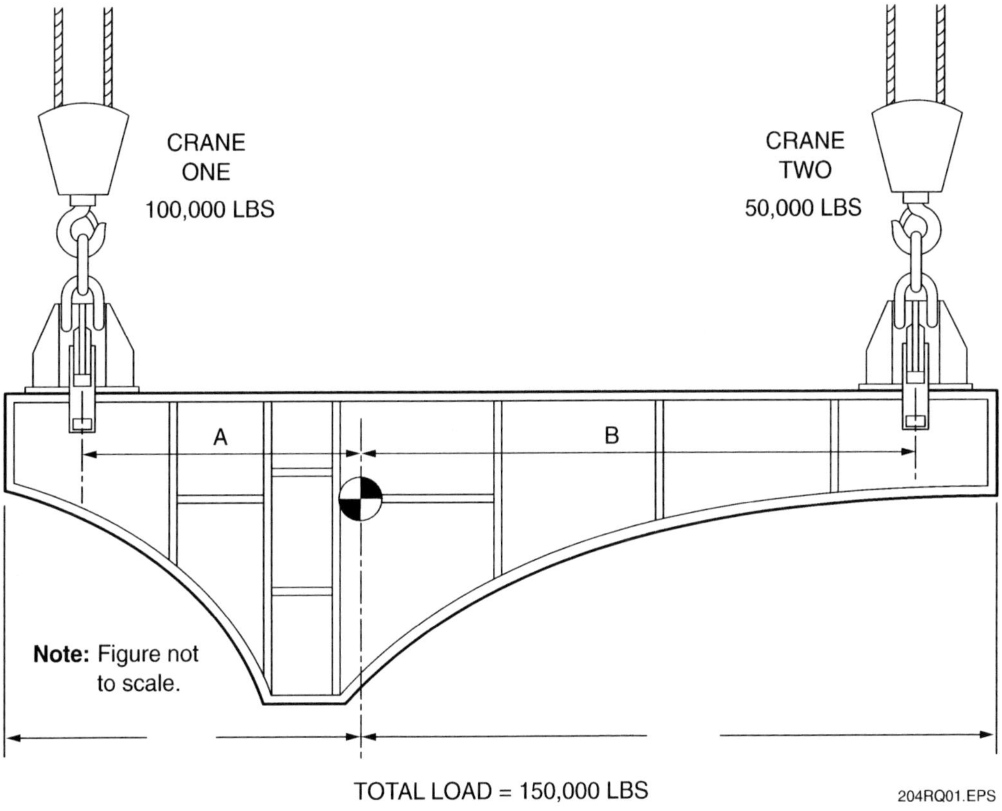

Figure 1

19. If A = 20 ft, then what is the value for B?
    a. 30 ft
    b. 35 ft
    c. 40 ft
    d. 45 ft

20. If A= 10 ft and B = 15 ft, then what is the actual load on Crane One?
    a. 57,000 lbs
    b. 60,000 lbs
    c. 90,000 lbs
    d. 150,000 lbs

21. In preparing for multiple-crane lifting, the load blocks of each crane have to be positioned relative to the load's center of gravity using the formula _____.
    a. (net capacity Crane One) × A = (net capacity Crane Two) × B
    b. (net capacity Crane One) + A = (net capacity Crane Two) + B
    c. (net capacity Crane Two) × A = (net capacity Crane One) × B
    d. (net capacity Crane One) – A = (net capacity Crane Two) – B

22. To determine the share of the load weight on Crane One that is positioned a distance of (A) feet from the load's center of gravity in a multiple-crane lift, the operator must use the formula _____.
    a. [B ÷ (A + B)] × load weight
    b. [B ÷ (A – B)] × load weight
    c. [A ÷ (A + B)] × load weight
    d. [A ÷ (A – B)] × load weight

23. Prior to planning a lift, a designated person must classify the lift into which of the following categories?
    a. Major, minor, critical
    b. Ordinary, critical, pre-engineered production
    c. Standard, critical
    d. General, special, critical

24. During critical lift operation, the operator must obey a _____ signal at all times.
    a. travel
    b. stop
    c. dog everything
    d. swing

25. A test lift is recommended before _____.
    a. any lift
    b. a lift involving a luffing jib boom
    c. a lift involving a 100-ton hook block assembly
    d. a critical lift

GLOSSARY

Trade Terms Introduced in This Module

Engineering: Practical application of the knowledge of physics, leverage, and gravity as they relate to lifting operations.

Equalizer: Any of the various devices or appliances for equalizing stresses, strains, and pressures.

Lift plan: A written plan that outlines all weight, rigging, and configuration considerations for a lift.

APPENDIX

Critical Lift Plan

APPENDIX A – CRITICAL LIFT PLAN

INDEX

1. OBJECTIVE

2. DEFINITIONS

2.1 Critical Lift
2.2 Minor Lift
2.3 Rigging
2.4 Lift Crane
2.5 Tail Crane

3. PROCEDURE

3.1 Responsibility for Critical Lift
3.2 Responsibility for Minor Lift
3.3 Special Lifting Requirements
3.4 Training

APPENDICES

Appendix 1 - Planned Lifting Procedure Form

PREPARED BY: ______________________
Engineering Representative

APPROVED BY: ______________________
Safety Representative

AUTHORIZED BY: ______________________
Project / Site Manager

DEVELOPED BY CRANE INSPECTION & CERTIFICATION BUREAU - 2004 1

204A01.EPS

CRITICAL LIFTS — *NOTES*

1 **PURPOSE**

1.1 Define responsibilities for personnel when crane lifts are made.

- To ensure the rigging and lift crane or cranes have adequate capacity for handling the weight of the device to be lifted.
- To ensure that the personnel responsible for the lift are qualified and competent personnel.
- To protect personnel and property.

2 **DEFINITIONS**

2.1 **Critical Lift** – As an accepted "General Industry Standard" The combined weight of the load, rigging and crane configuration exceeds 75% - 85% of the capacity of the crane and or there are complications as a result of load configuration, clearance, danger to personnel, power transmission equipment, critical service equipment or hazardous materials.

- The weight of the load is 25 tons or greater and exceeds 50' in length, height, and width. (Detailed Lift Plan required)
- Lifts involving more than one crane to handle a common load.

2.2 **Minor Lift** - The combined weight of the device, rigging and crane configuration is less than 80% of the crane capacity and there are no compilations as above.

- The lift weight is less than 25 tons and 50' in length.
- Lift is single crane.

2.3 **Rigging** – Any component part used to connect the device to be lifted by the crane; i.e., choker, sling, shackles, jib, spreader bars, snatch block, etc.

2.4 **Lift Crane** – The primary crane used to ultimately support the total weight of the device.

2.5 **Tail Crane** - The equipment used to tail or maneuver the device while the lift crane supports the device.

DEVELOPED BY CRANE INSPECTION & CERTIFICATION BUREAU - 2004

2

204A02.EPS

3 PROCEDURE

3.1 Responsibility for Critical Lifts

3.1.1 The Rigging Superintendent making the lift is responsible for planning and executing the lift. The superintendent shall complete the attached lifting procedure and acquire all approval signatures prior to the lift. The original "approved" procedure is to be submitted to the organization's Engineering Representative and Safety director for record and one copy retained by the Construction Rigging Superintendent. The Rigging Superintendent must include the organization's Engineering Representative and Safety Director in developing the lift plan to ensure full consideration of existing conditions and circumstances. Any lifts exceeding 85% must have Project / Site Manager 's signature.

The planning should include:

- Determining the maximum weight of the device to be lifted.
- Calculating the combined weight of the device, rigging accessories, load blocks and crane jibs.
- Calculating the sling and rigging hardware capacities with consideration to load angle factors and stresses on all slings, hardware and lift points.
- Determining the required boom angle and swing radius.
- Selecting the proper crane for the lift and the tail crane when required.
- Inspecting the lift site to determine the soil condition to decide whether matting and / or soil stabilization is needed.
- Determine the area to be restricted during lift.

204A03.EPS

3.1.2 After a careful review of the job site, the hoisting equipment and critical lift plan, the Rigging Supervisor will record the project data on the Planned Lifting Procedure Form. The crane operator will sign and date the completed form. In addition to his signature, the lifting procedure will be reviewed and signed by others as indicated on the lift form.

3.1.3 The crane operator has the authority and responsibility to ensure the lift is proper and in accordance with established local procedures, as well as OSHA and ANSI standards.

3.2 Responsibility for Minor Lifts

3.2.1 The Rigging Superintendent or competent person making the lift is responsible for completing the "Lifting Procedure" planning / executing the lift and submitting the lifting procedure to the Engineering Representative for record. Note: Repetitive minor lifts need only one "Lifting Procedure" completed per day of lifting activities; i.e., multiple pieces of structural steel and pipe installation. However, if the crane is repositioned, another "Lifting Procedure" should be completed before lifting is resumed.

- Should hoisting activities become routinely cycle duty procedures and located in remote locations the Engineering Representative or Project Superintendent may waive the completion of a lift form after review.

The Planning Should Include:

- Determining the maximum weight of the device to be lifted.
- Selecting the rigging based upon the maximum weight and calculate sling and hardware stresses.
- Calculating the combined weight of the device, rigging, load blocks and crane jibs.
- Determining the center of gravity of the load.

204A04.EPS

- Determining the required boom length, boom angle and swing radius at the pick point and the point the load is to be placed.
- Selecting the proper crane for the lift and the tail crane when required.
- Inspecting the lift site to determine the soil condition to decide whether matting and/or soil stabilization is needed.
- Identify the location of all utilities.
- Determine the area to be restricted to public access during lift.
- Determine the swing area.

3.2.2 After the careful review of the job site, hoisting equipment and the lift plan is completed, the Rigging Supervisor will record the project data on a Planned Lifting Procedure Form. The crane operator will sign and date the completed form. In addition to his signature, the Lifting Procedure will be reviewed and signed by the rigger.

3.3 Special Lifting Requirements

Whenever lifting devices which are smooth surfaced and lightweight; i.e., fiberglass or aluminum vessels, with canvas or nylon slings a strongback and / or bracing sling shall be used to prevent the lifting slings from rendering. Rendering (or slippage) of the main lifting slings can occur if the resistant force of friction between the sling and vessel wall is exceeded due to lift movement. Therefore, the strongback and / or bracing slings must be properly attached to prevent the main lifting slings from rendering in any position during the entire lift.

3.4 Training

3.4.1 Each person whose job requires them to operate cranes shall receive the following training:

- The recognition of the hazards associated with lifting equipment.

204A05.EPS

- The fundamentals of rigging practices and the understanding of load capacity charts as they pertain to crane configurations.
- The method and means of calculating load capacities.
- Attend a training course in mobile crane operations and rigging considerations.

3.4.2 Each person who is involved in rigging shall be required to participate in a rigging training program and tested to become competent in the selection of rigging when a lift is performed.

3.4.3 All trades personnel and supervision involved in lifting shall be instructed about this procedure and the requirements.

204A06.EPS

3.4.4 PLANNED LIFTING PROCEDURE FORM

1. DATE OF FORM PREPARATION:	
2. PROJECT NUMBER:	
3. PROJECT TITLE / LOCATION:	
4. PERMIT REQUIRED: Yes / No	
5. PREPARED BY:	
6. LOAD DESCRIPTION / PHYSICAL SIZE:	
• Equipment Name	
• I.D. / Tag Number	
• Physical Dimensions	
• Equipment Orientation V or H	
• Lift Lug Orientation Degrees	
• Hoisted load / Device Age	
• Physical Condition	
• NDE Required Y / N	
7. LIFTING AND TAILING ATTACHMENTS APPROVED: Y / N	
8. LOAD ELEVATION:	
9. LOAD WEIGHT:	

204A07.EPS

APPLICATION - LIFT CRANE

10. CRANE(S) TYPE:

11. CRANE CONFIGURATION:

12. PARTS OF WIRE ROPE: Wire Rope Breaking Strength

- Single Line Pull Wire Rope SWL

13. MAXIMUM RADIUS:

14. BOOM ANGLE:

15. BOOM CLEARANCE CHECK REQUIRED: Yes No

16. TAIL SWING CLEARANCE VERIFIED: Yes No

17. CHART CAPACITY:

18. CRANE RIGGING:	SIZE	CAPACITY	CONDITION
• Choker(s)			
• Sling(s)			
• Shackle(s)			
• Spreader Bar(s)			
• Other			

19. CRANE RIGGING COMPONENT WEIGHT:

- Jib Weight
- Ball Weight
- Load Block Weight
- Suspended Load Line Weight
- Shackle(s) Weight
- Choker(s) Weight
- Sling(s) Weight
- Spreader Bar(s) Weight
- Other Lifting Attachment Weights
- Total Rigging Weight

20. GROSS LOAD:

21. % OF CHART:USED

DEVELOPED BY CRANE INSPECTION & CERTIFICATION BUREAU - 2004

8

204A08.EPS

APPLICATION - TAIL CRANE

10. CRANE(S) TYPE:

11. CRANE CONFIGURATION:
12. PARTS OF WIRE ROPE: Wire Rope Breaking Strength
- Single Line Pull Wire Rope SWL

13. MAXIMUM RADIUS:
14. BOOM ANGLE:
15. BOOM CLEARANCE CHECK REQUIRED: Yes No
16. TAIL SWING CLEARANCE VERIFIED: Yes No
17. CHART CAPACITY

18. CRANE RIGGING	SIZE	CAPACITY	CONDITION
• Choker(s)			
• Sling(s)			
• Shackle(s)			
• Spreader Bar(s)			
• Other			

19. CRANE RIGGING COMPONENT WEIGHT:
- Jib Weight
- Ball Weight
- Load Block Weight
- Suspended Load Line Weight
- Shackle(s) Weight
- Choker(s) Weight
- Sling(s) Weight
- Spreader Bar(s) Weight
- Other Lifting Attachment Weights
- Total Rigging Weight

20. GROSS LOAD:
21. % OF CHART USED:
22. RIGGING SAFETY DESIGN FACTOR 5 to 1 Yes No

DEVELOPED BY CRANE INSPECTION & CERTIFICATION BUREAU - 2004

9

204A09.EPS

CRITICAL LIFTS

NOTES

Engineering Certification Spreader Bar (Required)

- Proof Test Certification of Chokers 1 1/4" &

Greater (Required)

23. SOIL TYPE / CONDITION:

24. MATS / CRIBBING Yes No

25. RIGGING INSPECTED: Yes No

26. TAG LINE:

27. HAZARDS

28. WEATHER:

29. DATE: TIME:

30. LIFT MEETING CONDUCTED: Yes No

31. APPROVALS:

Rigging Supervisor: ______________________________

Crane Operator. ________________________________

Rigger / Signalman. ______________________________

Client Representative (If Required):

Safety Representative (If Required):

Engineering Representative (If Required)

Project Manager (If Required)

Communications: What form of communications will be used and by whom?

DEVELOPED BY CRANE INSPECTION & CERTIFICATION BUREAU - 2004

10

204A10.EPS

INSTRUCTIONS FOR FILLING OUT PLANNED LIFTING PROCEDURE FORM

1. **Date of form preparation**.
2. **PROJECT NUMBER:** Enter the project number that has been assigned to this work activity.
3. **PROJECT TITLE/LOCATION:** Enter the name of the project, plant site or construction site of the lift.
4. **PERMIT REQUIRED (y/n):** Indicate if a permit will be required. Certain plant safe work procedures require a permit when making certain lifts over in-service facilities or lifts made near power lines or guy lines or exceeding 75% of load crane capacity chart.
5. **PREPARED BY:** Enter name of person responsible for research and preparation of this form.
6. **LOAD DESCRIPTION / PHYSICAL SIZE:** Describe the load to be lifted including name of the equipment and item number and or tag number. Enter physical dimensions of load to be lifted. Identify equipment orientation with a "V" or "H", "V" for vertical and "H" for horizontal. Identify age of existing equipment / device to be lifted, physical condition and / or if NDE is required on item to be lifted prior to lift.
7. **LIFTING / TAILING ATTACHMENTS APPROVED (y/n):** Indicate if lifting attachments are approved by, the equipment engineering firm.
8. **LOAD ELEVATION:** Indicate the maximum and minimum elevation in feet that the load will travel Enter the maximum radius the load will achieve during the lift cycle of pick, swing & set.
9. **LOAD WEIGHT:** Determine the weight of the load utilizing certified engineer drawings, vendor data or operations manual. Only use load weight information, which has been documented and certified, or calculated by a Mechanical or Vessel P.E.
10. **CRANE(S) TYPE:** Enter the type, model and rated capacity of lift equipment to be used for lifting
11. **CRANE CONFIGURATION:** Indicate the length of the main boom and the length of jib (if equipped) that will be in the lift & tail crane(s) at time of lift.
12. **PARTS OF WIRE ROPE / SINGLE LINE PULL:** Enter the number of parts in the load handling line during the lift. Enter the allowable winch single line pull in lbs.

204A11.EPS

13. **MAXIMUM RADIUS:** Enter the maximum working radius the crane / load will achieve during the lift cycle of pick, swing & set.

14. **BOOM ANGLE:** Indicate the cranes' boom angle at the initial pick point / radius and final set point/radius.

15. **BOOM ANGLE:** Indicate the cranes' boom angle at the initial pick point / radius and final set point / radius.

16. **TAIL SWING CLEARANCE VERIFIED (y/n):** Indicate if the tail swing clearance has been verified. When it is required, the sketch / diagram verifying the tail swing clearance should be attached to the PCR Planned Crane Lifting Procedure Form.

17. **CHART CAPACITY:** Indicate the capacity of the crane(s) from the manufacturer's capacity chart with the parameters as indicated in items (8) thru (9).

18. **CRANE RIGGING:** Enter the size of chokers and shackles used for the lift and their physical condition.

19. **RIGGING COMPONENT WEIGHTS:** Enter the manufacturer's effective weight for the jib, load block, suspended load line, ball, suspended whip / load line, shackles, slings, spreader bars and any other lifting attachments suspended under the boom point.

20. **GROSS LOAD:** Indicate the total weight of the load by adding (9) and (19).

21. **% OF CAPACITY CHART USED:** Indicate the percentage of the crane's chart capacity to be utilized for the lift. If load / capacity percentage exceeds 75%, the lift should not be made without review of PCR Engineering Representative and Project / Site Manager. "Net load divided by net capacity = % Capacity Used "

22. **RIGGING SAFETY FACTOR 5 TO 1 (y/n):** Verify that the rigging equipment (shackles, chokers, etc.) used to perform the lift have a 5 to 1 Safety Factor.

    - Engineering Certification on Spreader Bars
    - Proof Test Certifications of Shackles and Slings

23. **SOIL CONDITION:** Indicate the type of soil in the area of the lift; loose, compact, or undisturbed earth. Consideration should be given to any adjacent excavation and its distance from the outriggers or crawler tracks.

24. **MATS REQUIRED:** Depending on the soil conditions determined above, indicate if any matting is required to properly protect any underground facilities and to secure the affected area under the

204A12.EPS

lifting equipment. Matting location, configuration and the number of mats should be identified on the rigging plans and diagrams.

25. **RIGGING REVIEWED (y/n):** Indicate if the rigging has been reviewed. This includes chokers, shackles, spreaders, load line, lifting lugs and any other lifting attachments required. Depending on the nature of the lift, it may be required that a lift plan and diagram should be developed and attached to the "Planned Crane Lifting Procedure Form." This may be required at the discretion of the Client Representative or the Rigging Supervisor.

26. **TAG LINE (y/n)** Indicate if a tag lines will be required.

27. **HAZARDS:** Indicate if there are any existing power lines or overhead obstructions in the area of the lift. In addition, indicate where there are any existing underground utilities or items buried in the crane set-up area. Identify the hazard (water, sewage, drainage, electrical, etc.) and its depth.

28. **WEATHER:** Indicate the weather conditions prior to and during the lift with emphasis on wind velocity and direction, rain probability, etc.

29. **DATE: TIME:** Indicate the date and approximate time of day that the lift will be made.

30. **LIFT MEETING CONDUCTED (y/n):** Indicate if a pre-lift meeting will be conducted with all involved parties. The client representative may request such a meeting.

31. **APPROVALS:** A minimum of four (4) signatures will be required on the lift form. These include the Crane Operator, Rigging Supervisor, Safety, Client Representative. The Client Representative, at his option, may request additional signatures including Equipment Supervisors, Safety Representatives, and the organization's Engineering Representative.

204A13.EPS

Figure Credits

Manitowoc Crane Group	204F01–204F03
Crane Inspection & Certification Bureau	Appendix

CONTREN® LEARNING SERIES — USER UPDATE

The NCCER makes every effort to keep these textbooks up-to-date and free of technical errors. We appreciate your help in this process. If you have an idea for improving this textbook, or if you find an error, a typographical mistake, or an inaccuracy in NCCER's Contren® textbooks, please write us, using this form or a photocopy. Be sure to include the exact module number, page number, a detailed description, and the correction, if applicable. Your input will be brought to the attention of the Technical Review Committee. Thank you for your assistance.

Instructors – If you found that additional materials were necessary in order to teach this module effectively, please let us know so that we may include them in the Equipment/Materials list in the Annotated Instructor's Guide.

Write: Product Development and Revision
National Center for Construction Education and Research
P.O. Box 141104, Gainesville, FL 32614-1104

Fax: 352-334-0932

E-mail: curriculum@nccer.org

Craft ____________________ Module Name ____________________

Copyright Date __________ Module Number __________ Page Number(s) __________

Description

____

____

____

____

(Optional) Correction

____

____

____

(Optional) Your Name and Address

____

____

____

Module 38205-06

Advanced Rigging Practices

COURSE MAP

This course map shows all of the modules in the second part of the *Rigging* curriculum, *Rigging*. The suggested training order begins at the bottom and proceeds up. Skill levels increase as you advance on the course map. The local Training Program Sponsor may adjust the training order.

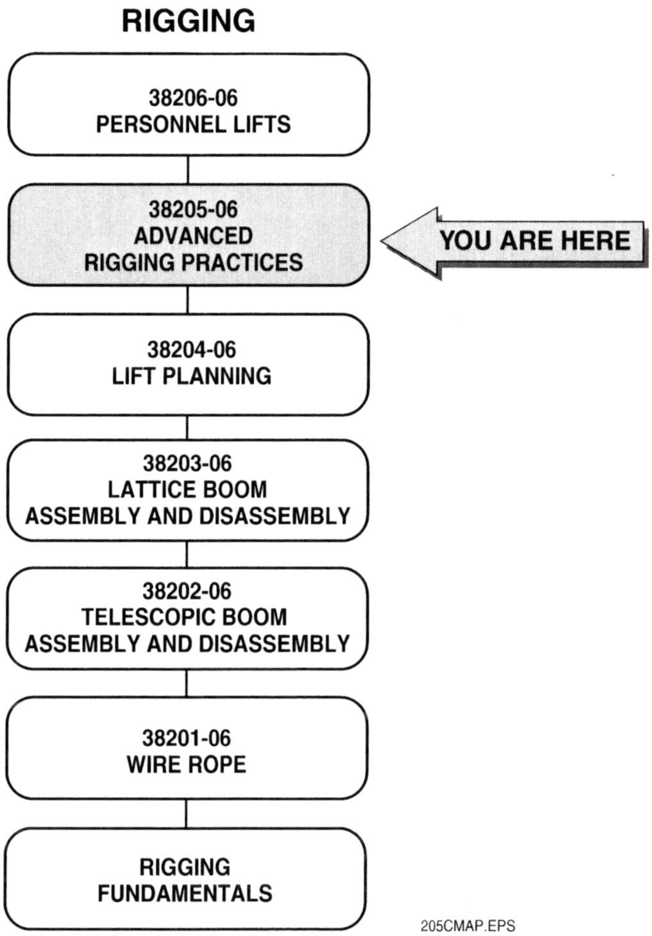

205CMAP.EPS

MODULE 38205-06 CONTENTS

1.0.0 INTRODUCTION 5.1

2.0.0 LOAD DYNAMICS 5.1

2.1.0 Rotational Forces or Moments 5.3

2.2.0 Crane Stability 5.5

2.2.1 Quadrants of Operation 5.6

2.2.2 Centers of Gravity and Forward Stability 5.7

2.2.3 Backward Stability 5.9

2.2.4 Noncentered Lifts and the Effects on Load Radius 5.9

2.2.5 Other Factors Affecting Stability 5.9

3.0.0 SPECIAL EQUIPMENT USED IN HEAVY RIGGING 5.9

3.1.0 Jacking 5.11

3.1.0 Grip Hoists 5.13

3.2.0 Skids 5.13

3.3.0 Rollers 5.13

3.4.0 Blocking 5.15

3.5.0 Cribbing 5.15

3.6.0 Inclined Planes 5.16

4.0.0 SLINGS 5.16

4.1.0 Sling Tensions 5.17

4.2.0 Bridle Hitches 5.19

4.3.0 Basket Hitches 5.19

4.4.0 Choker Hitches 5.20

5.0.0 USING BEAMS 5.21

5.1.0 Adjustable Beams 5.22

5.2.0 Equalizer Beams 5.23

6.0.0 RIGGING REBAR BUNDLES 5.23

6.1.0 Unloading Procedure 5.23

6.2.0 Hoisting Equipment 5.23

7.0.0 SUBMERGED LIFTS 5.24

8.0.0 COMMUNICATION 5.24

SUMMARY 5.25

REVIEW QUESTIONS 5.26

GLOSSARY 5.27

APPENDIX A, Installation of a Tube Bundle in a Heat Exchanger 5.28

APPENDIX B, Removing a Horizontal Heat Exchanger Using an Extractor 5.31

APPENDIX C, Vertical Vessel Rigging 5.33

REFERENCES & ACKNOWLEDGMENTS 5.35

Figures

Figure 1 Horizontal rigging application 5.2
Figure 2 Multi-crane lift 5.2
Figure 3 Heavy-lift cranes 5.3
Figure 4 Loads in balance 5.4
Figure 5 Unbalanced loads 5.4
Figure 6 A theoretically balanced simple crane 5.5
Figure 7 Operational quadrants 5.6
Figure 8 Stability example one 5.7
Figure 9 Stability example two 5.8
Figure 10 Boom positions affecting stability and tipping axis ...5.8
Figure 11 Stability conditions affected by wind 5.10
Figure 12 Hydraulic jacks 5.11
Figure 13 Hydraulic cylinders 5.11
Figure 14 Hydraulic pumps 5.11
Figure 15 Jacking systems 5.12
Figure 16 Grip hoist 5.13
Figure 17 Use of rollers 5.14
Figure 18 Cribbing 5.15
Figure 19 Inclined plane 5.16
Figure 20 Sling measurement 5.16
Figure 21 Sling angle factors 5.17
Figure 22 Bridle hitch arrangement 5.19
Figure 23 Asymmetrical bridle sling stress arrangement and formula 5.20
Figure 24 Choker hitch arrangement 5.20
Figure 25 Choker hitch derating factors 5.21
Figure 26 Spreader beam application 5.21
Figure 27 Multiple spreader beam application 5.22
Figure 28 Adjustable lifting beam 5.22
Figure 29 Use of a spreader beam to rig rebar 5.23
Figure 30 Proper way to use two slings to lift a bundle 5.24

Tables

Table 1 Example of a Sling Capacity Chart 5.18

MODULE 38205-06

Advanced Rigging Practices

Objectives

1. Explain how the center of gravity of the load affects the rigging.
2. Explain how the weight of the load and the position of the crane boom affect the load capacity of the crane.
3. Describe how jacks, tuggers, hoists, skids, and rollers are used to move loads laterally.
4. Explain how cribbing and blocking are used to support loads.
5. Given a particular load, select the appropriate sling for a lift.
6. Demonstrate the proper use of hand, voice, and audible signals.
7. Select the appropriate spreader bar or equalizer beam for a given load.
8. Demonstrate the ability to place the hook over the center of gravity for a non-symmetrical load.
9. Perform or simulate a blind lift using voice communication.

Prerequisites

Before you begin this module, it is recommended that you successfully complete *Rigging Fundamentals* and *Rigging*, Modules 38201-06 through 38204-06.

Required Trainee Materials

1. Pencil and paper
2. Appropriate personal protective equipment

1.0.0 ◆ INTRODUCTION

The previous level covered basic rigging equipment and methods that could be applied by construction workers who may be called upon to assist in general rigging tasks as part of their craft work. This module covers more complex rigging tasks that require the advanced skills of a rigging specialist.

Rigging projects come in all shapes, weights, and movement directions. Rigging activities may involve jacks, rollers, winches, cranes, or a combination of rigging gear. It is a good practice to develop a plan prior to beginning the work.

At various industrial sites, including power plants and refineries, riggers are required to move huge items of equipment and other special loads within tight spaces. In many cases, it is necessary to move the load laterally before it can be lowered. An example is the removal of a heat exchanger (*Figure 1*).

In some instances it will take more than one crane to safely lift or erect a structure such as a tank or pressure vessel (*Figure 2*). This is very specialized rigging that requires skilled crane operators and riggers working together.

Although large lifts are usually engineered, the rigger must interpret and implement the engineer's design and make sure that the right equipment and rigging methods are being used.

NOTE

The Appendixes to this module contain case studies of some complex rigging problems. These cases will give you some insight into the planning and implementation for the rigging of large, unwieldy loads.

2.0.0 ◆ LOAD DYNAMICS

As discussed in *Rigging Fundamentals*, a mobile crane (*Figure 3*) is a machine that lifts and moves loads within its designed capabilities.

205F01.EPS

Figure 1 ◆ Horizontal rigging application.

205F02.EPS

Figure 2 ◆ Multi-crane lift.

205F03.EPS

Figure 3 ◆ Heavy-lift cranes.

Cranes operate according to the principles of **leverage** and stability. Leverage, as previously described, is the force applied to a **lever** to move a load. A crane uses this principle to lift loads. This concept can be visualized as a very large and sophisticated teeter-totter (or seesaw).

Look at a simple teeter-totter, which operates by having two opposing leverage forces counteract each other. This opposition takes place across a tipping fulcrum (**tipping axis**). As the weights at each end of the teeter-totter become unbalanced, the heavier side applies leverage to the lighter side. The result is that the lighter side will be lifted up until it is stopped or until the leverage forces become equal, and the lever is again balanced.

2.1.0 Rotational Forces or Moments

In *Figure 4*, equal weights are placed at equal distances from the fulcrum. This maintains the lever in a state of balance because the clockwise (CW) and counterclockwise (CCW) rotational forces at the fulcrum are equal. The rotational forces are a product of the **center of gravity** of the weights and their distances from the fulcrum and are defined as moments. In *Figure 5*, the distances are again equal, while the weights are unequal. If the tipping fulcrum remains at the same position, the heavier side will cause the lighter side to rise. This movement will only stop when the rotational forces applied to both sides of the lever are again in balance. To balance the rotational forces in *Figure 5*, the heavier weight must be supported (made lighter) or more force must be applied to the lighter side. In *Figure 5*, the heavier weight side will be reduced when it contacts the ground. This equalizes the forces between the sides, again balancing the lever.

Another way to balance the rotational forces between the two sides is by moving the tipping fulcrum toward the heavier side. This method shortens the distance between the center of gravity and the tipping fulcrum on the heavier side, while lengthening it on the lighter side. This results in a proportional change in the moments applied to both sides.

To calculate the rotational forces of a lever about a pivot point, use the following simple formula:

The weight (W) times the distance (X) from the center of gravity of the weight to the pivot point equals a rotational force, or moment, expressed in foot-pounds of torque.

Figure 4 shows weights in balance. To be balanced, the CW and CCW forces must be equal, as demonstrated in an application of the previous formula.

W1 × X1 = W2 × X2

500 lbs × 10 ft = 500 lbs × 10 ft

5,000 ft-lbs = 5,000 ft-lbs

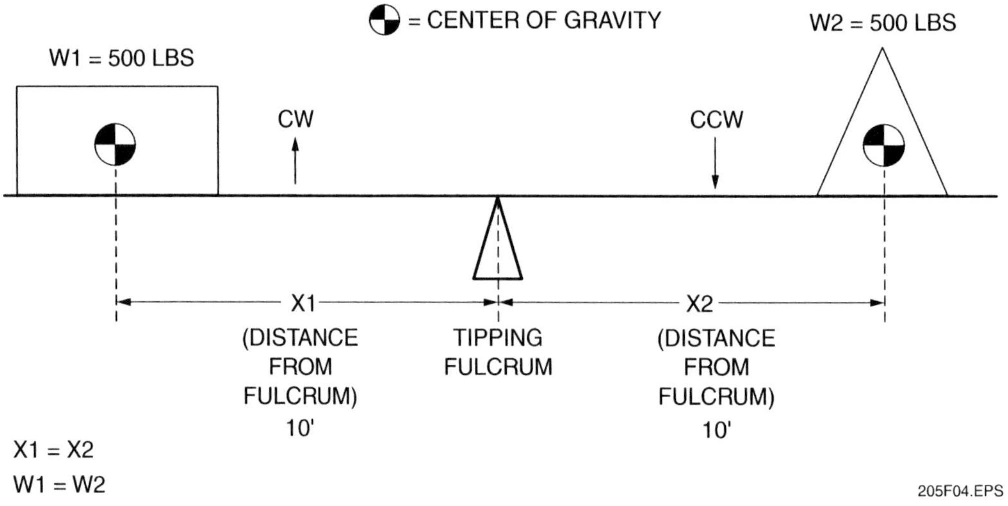

Figure 4 ◆ Loads in balance.

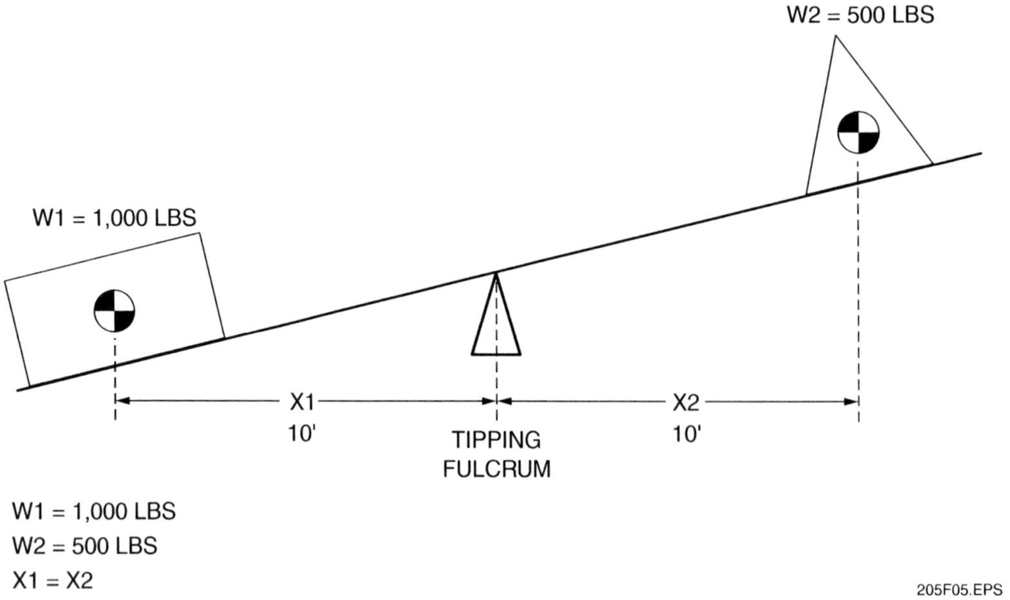

Figure 5 ◆ Unbalanced loads.

Figure 5 shows an unbalanced load. In this example, an unbalanced condition with a rotational force of 5,000 ft-lbs exists counterclockwise around the tipping fulcrum. This rotational force is shown in the following calculation:

$W1 \times X1 = W2 \times X2$

$1{,}000 \text{ lbs} \times 10 \text{ ft} = 500 \text{ lbs} \times 10 \text{ ft}$

$10{,}000 \text{ ft-lbs} \neq 5{,}000 \text{ ft-lbs}$

$10{,}000 \text{ ft-lbs} - 5{,}000 \text{ ft-lbs} = 5{,}000 \text{ ft-lbs}$ rotational force moment for $W1 \times X1$

To balance the lever in *Figure 5*, the weight of sides would have to change, or the distances would have to change. Assume the weights remain steady; the only change will be in the distances from the fulcrum. To calculate the distance change required to restore the balance, X1 can be shortened and/or X2 can be lengthened. To keep X1 the same and solve for a lengthened X2, calculate the following:

$W1 \times X1 = W2 \times X2$

$$\frac{W1 \times X1}{W2} = X2$$

$$\frac{1{,}000 \text{ lbs} \times 10 \text{ ft}}{500 \text{ lbs}} = X2$$

$X2 = 20 \text{ ft}$

To solve for a shortened X1 with X2 unchanged, calculate as follows:

$W1 \times X1 = W2 \times X2$

$$\frac{W2 \times X2}{W1} = X1$$

$$\frac{500 \text{ lbs} \times 10 \text{ ft}}{1{,}000 \text{ lbs}} = X1$$

$X1 = 5 \text{ ft}$

By changing the shape of the teeter-totter, as in *Figure 6*, the outline begins to take on the shape of a simple crane. This illustrates that a crane uses nothing more than the principle of leverage, as defined by rotational forces, to lift a load.

The number of variables that are introduced into the leverage calculation make crane operation more complicated and dynamic than that of a simple teeter-totter. During a lift with a crane, the distance from the crane's center of gravity to the tipping axis will change as the crane is rotated. The load's center of gravity to the tipping axis will change when the load is boomed in or out. As the boom is loaded, it will deflect, causing changes as well. Several changes occur simultaneously while booming up and swinging the load.

A mobile crane is stable when its leverage on the load is greater than the load's leverage, which is also called a load moment, on the crane. A crane's stability decreases as the load radius increases or as heavier loads are lifted. If the leverage of the load exceeds the leverage of the crane, the crane will tip. If both leverages are equal, the crane's stability is in jeopardy. Theoretically, in an equal leverage situation, the crane would tip over by simply being pushed up in an area near the counterweights at the rear of the crane. While viewing *Figure 6*, the following calculation illustrates this principle:

$W1 \times X1 = W2 \times X2$

$60{,}000 \text{ lbs} \times 20 \text{ ft} = 30{,}000 \text{ lbs} \times 40 \text{ ft}$

$1{,}200{,}000 \text{ ft-lbs} = 1{,}200{,}000 \text{ ft-lbs}$

In this situation, the crane could become unstable at any moment. It is imperative that the amount of rotational force on the crane side of the equation always be greater (unbalanced) than that on the side of the load. This ensures that the crane will not tip when the load is raised. When cranes are designed, the gross load that they are capable of lifting at a given load radius is specified on a load chart. This chart takes into consideration the configuration of the crane and the moving centers of gravity while performing a lift.

2.2.0 Crane Stability

Crane stability depends on several factors. These factors include the quadrant of operation of the crane, forward and backward stability, noncentered lifts, and environmental factors.

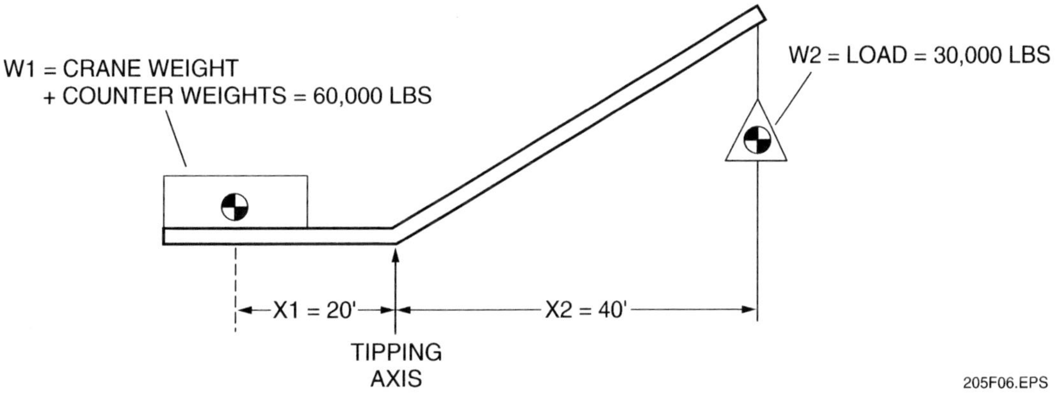

Figure 6 ◆ A theoretically balanced simple crane.

2.2.1 Quadrants of Operation

As a crane swings to various quadrants, the distance that the load is positioned from the tipping axis changes dramatically. This is because cranes are typically built to be longer than they are wide to allow them to fit on a roadway or trailer for transportation from job to job. This rectangular-shaped footprint tends to result in more stability of the crane from front to rear than from side to side.

Manufacturers define the operating quadrants for each of the cranes that they produce. *Figure 7* shows one example.

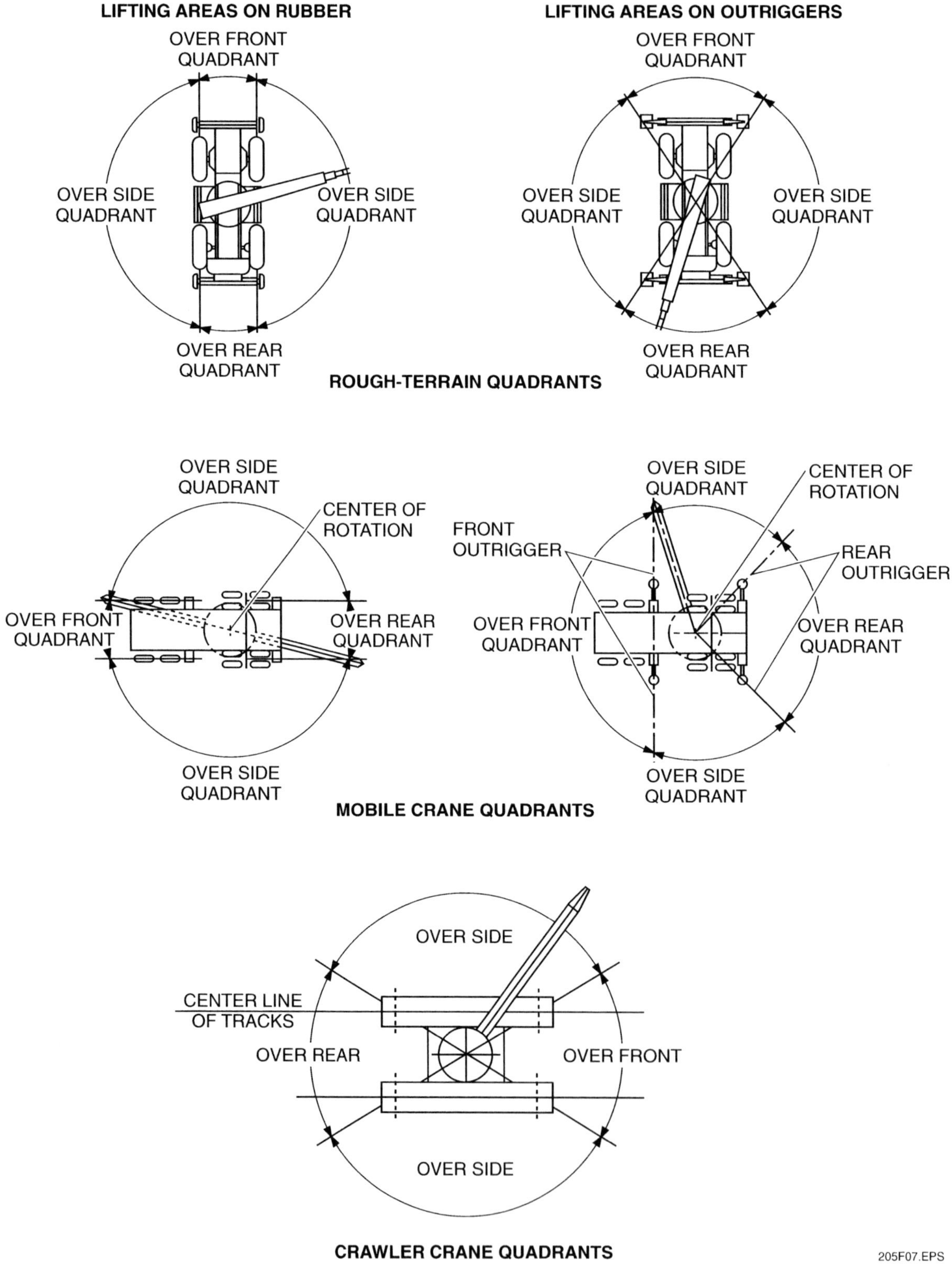

Figure 7 ◆ Operational quadrants.

Along with the definition of the quadrants, the manufacturer also identifies the authorized quadrants of operation and specifies any limitations while operating in those areas. The quadrants of operation are defined by referencing the boom position in relation to its position over the carrier body. The quadrants are as follows:

- *Over-the-front* – An area located between the front outriggers or crawlers when facing in the forward direction of travel.
- *Over-the-rear* – The area between the rear outriggers or crawlers when facing the rearward direction of travel.
- *Over-the-side* – A position located between the left- or right-side outriggers or over the crawler.
- *360-degree* – The circular swing area that would encompass all of the previously discussed quadrants.

2.2.2 Centers of Gravity and Forward Stability

The movements of the various centers of gravity in relation to the tipping fulcrum (tipping axis) while operating a crane affect the crane's forward stability. The most important things to understand are the crane's combined center of gravity and the load's center of gravity. The center of gravity of the crane is a combination of the upper works (superstructure), base section, and counterweight centers of gravity. Changes that cause movement in a crane's center of gravity are as follows: reconfiguring the counterweights, rotating the upper works, extending or retracting the outriggers, and extending or retracting the crawlers. The movements that affect the load's center of gravity in relation to the tipping axis are as follows: extending, retracting, raising, and lowering the boom or changing a luffing jib angle.

The distribution of the forces on either side of the tipping axis will determine the stability of the crane. As the crane's combined center of gravity moves toward the tipping axis, the calculated leverage applied by the crane is reduced. This situation also occurs if the load is moved further from the tipping axis. Moving the load further from the tipping axis increases the leverage applied by the load. Both of these conditions cause the leverage forces to equalize, producing an unstable crane condition. The following mathematical examples show this relationship.

Example One:

Assume a load is being lifted over the front of a crane. The weights of the load and the crane are shown in the diagram (*Figure 8*); their distances from the tipping axis are also shown. The forces applied on each side of the tipping axis are determined by multiplying the weight of the object by its distance from the tipping axis.

Leverage of the crane:

4,000 lbs (crane weight) × 8 ft (distance of crane CG from tipping axis) = 32,000 ft-lbs (leverage of crane)

Leverage of the load (load moment):

500 lbs (load weight) × 20 ft (distance of crane CG from tipping axis) = 10,000 ft-lbs (leverage of load)

The crane's leverage (32,000 ft-lbs) is much greater than the load moment (10,000 ft-lbs). The crane will be stable.

Example Two:

Assume that the crane has been rotated to lift the load from the over-the-side position. The crane and load weights remain the same, as shown in *Figure 9*, but the distances from the tipping axis have changed.

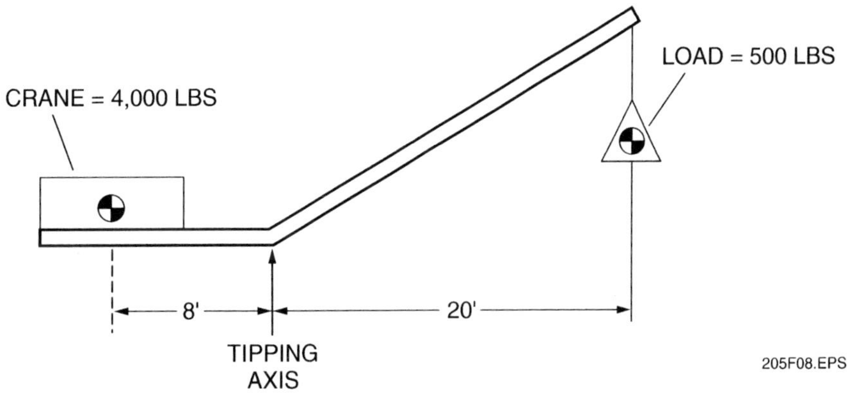

Figure 8 ◆ Stability example one.

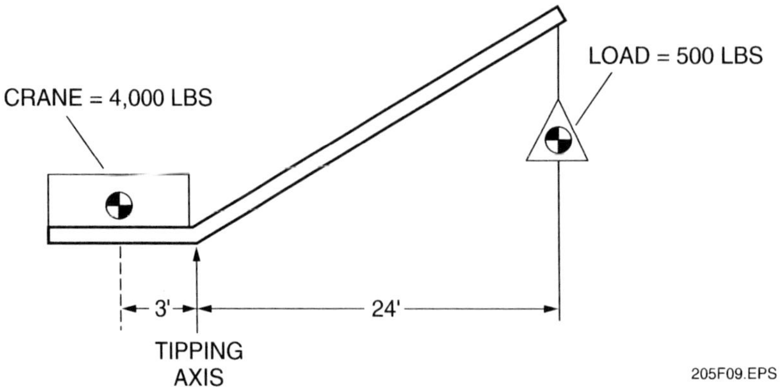

Figure 9 ◆ Stability example two.

The forces applied on each side of the tipping axis are determined by multiplying the weight of the object by its distance from the tipping axis.

Leverage of the crane:

4,000 lbs (crane weight) × 3 ft (distance of crane CG from tipping axis) = 12,000 ft-lbs (leverage of crane)

Leverage of the load (load moment):

500 lbs (load weight) × 24 ft (distance of crane CG from tipping axis) = 12,000 ft-lbs (leverage of load)

The crane's leverage (12,000 ft-lbs) is equal to the load moment (12,000 ft-lbs). This situation should not be attempted, as the crane will be very unstable. Also note that the tipping axis on the crane moved as the crane was rotated. The new tipping axis remained between the load and the crane's center of rotation. Moving the crane's boom from one operating quadrant to another can have a dramatic effect on crane stability. When an operator takes action that reduces the stabilizing force on the crane, it can result in a crane tipping accident. This effect is illustrated in *Figure 10*.

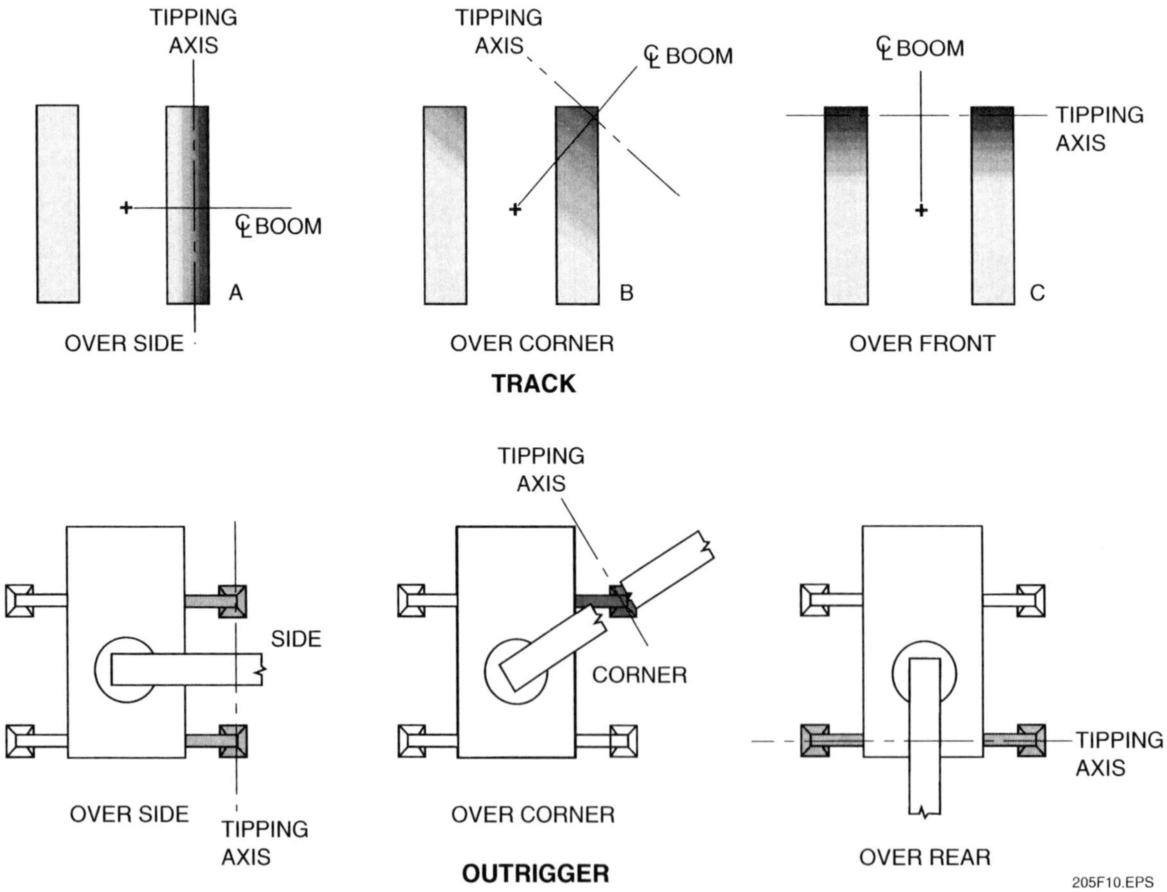

Figure 10 ◆ Boom positions affecting stability and tipping axis.

2.2.3 *Backward Stability*

In addition to forward stability, the operator must also be concerned with reverse, or backward, stability. The dynamics are only slightly different from those of forward stability. The concern is to keep the crane's center of gravity from going past the backward tipping axis. If this situation occurs, the crane could tip over backwards.

Several factors can contribute to backward tipping. One is the sudden movement of the boom or load. This movement occurs by suddenly stopping the boom while raising or by suddenly releasing the load. The deflection in the boom toward the crane momentarily shifts the center of gravity past the tipping axis, resulting in tipping.

Another factor is the orientation of the crane and boom to the ground. If the operator places a long boom in an extreme upright position and travels up an incline, the crane's center of gravity moves past the rear tipping axis. The high boom angle and incline combine to decrease the distance between the crane's center of gravity and tipping axis. This reduces the stabilizing forces, resulting in a crane tipping accident.

2.2.4 *Noncentered Lifts and the Effects on Load Radius*

Noncentered lifts affect a crane's stability and structural integrity. If the load is not picked up with the hook and boom centered directly over the load's center of gravity, side loading of the boom and/or an increase in load **radius** will occur. As the crane picks up the load, it will result in additional load weight, added by the friction of dragging the load. This situation also results in load movement, causing the tipping axis to continuously move to a point between the load and the rotating section of the crane. The tipping axis movement will continue until the load and hook blocks have reached a point directly below the boom. The dynamic that worsens this condition is the **pendulum effect**, which occurs when the load clears the ground. As the load clears the ground, its center of gravity will move to a point opposite from where it started. This pendulum effect creates a very unstable condition, resulting in wide variations in boom side loading and/or the measured load radius. This will overload the crane, causing either a structural failure or a tipping incident.

2.2.5 *Other Factors Affecting Stability*

One of the most overlooked stability factors affecting cranes is the wind (see *Figure 11*). High wind speeds can dramatically affect a crane and its load. Almost all crane manufacturers specify in the load chart that chart ratings must be reduced under windy conditions, and they may also recommend a shutdown at a certain wind velocity. In almost all cases, when the wind speed exceeds 30 mph, it is advisable to stop operations.

Wind affects both the crane and the load, reducing the rated capacity of the crane. Operators must never make a full-capacity lift if it is windy. Use a great deal of discretion even when lifting under moderate wind conditions of 20 mph. It is advisable to avoid handling loads that present large wind-catching surfaces. The result could be loss of control of the load and crane even though the weight of the load is within the normal capacity of the crane.

A 20 mph wind exerts a force of only 1⅛ lb/ft^2 on a flat-surfaced load (the force on a 4 ft by 8 ft sheet of plywood = 36 lbs), so only loads having very large sail areas would require derating the crane capacity. At 30 mph, however, the wind exerts a force of 2½ lb/ft^2 on a flat surface area (80 lbs on a sheet of 4 ft by 8 ft plywood). This wind force on the load at 30 mph is enough to cause non-vertical hoist lines and loads that are very difficult to control.

Forward stability is the critical consideration when the wind is coming from behind the boom. It applies a force to the boom and to the load that adds to the overturning moment of the crane. This has the same effect as adding load to the hook. Backward stability is the critical factor when the wind is from the front, particularly when the boom is at or approaching the maximum boom angle. The wind forces on the boom reduce the forward tipping effect normally provided by the boom. Without it, the crane can tip backward; if the wind speed is sufficiently high, the loads created can topple the boom.

Another equally dangerous dynamic effect is known as impact loading. Impact loading is when the sudden release of a load causes a rapid change in effective load weight. When this happens, the crane will recoil initially, and then the full weight of the load will suddenly be applied to the hook. The result can be overturning or structural failure of the crane. Similar to impact loading is rapid **acceleration** or deceleration of a load. A rapid hoist acceleration will produce a hook load higher than the actual load weight. If there is a sudden release of a load, it can cause a boom to recoil and topple over backwards. The crane operator should remember to ease into the load when hoisting. When lowering, the load should be placed gently to allow boom deflection and pendant stretch to gradually return to normal.

3.0.0 ◆ SPECIAL EQUIPMENT USED IN HEAVY RIGGING

Although cranes are used most often, other equipment such as jacks, tuggers, grip hoists, skids, rollers, and inclined planes are also used to position equipment, especially for horizontal movement.

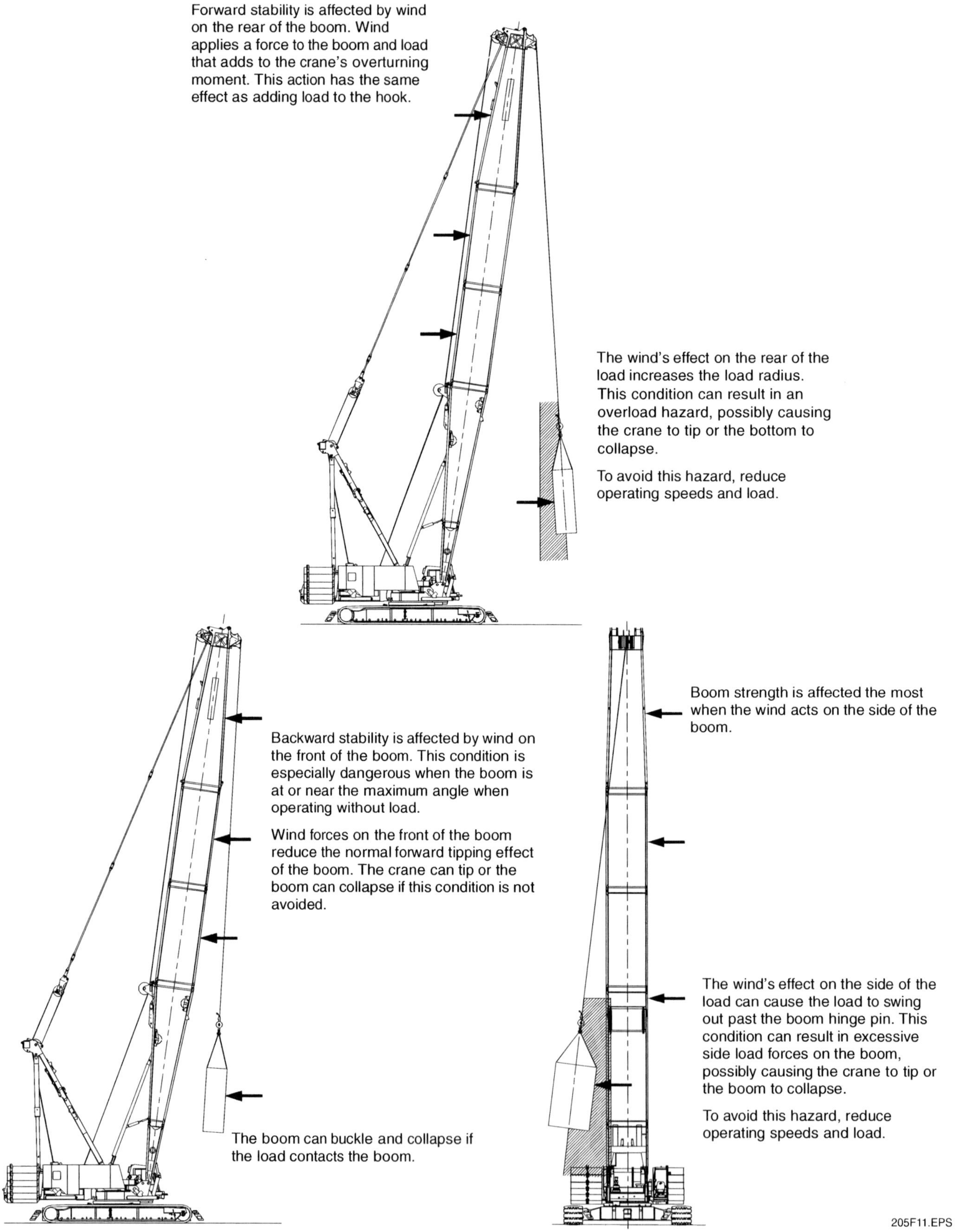

Figure 11 ◆ Stability conditions affected by wind.

3.1.0 Jacking

Jacks are used to raise or lower a load. Hydraulics jacks (*Figure 12*), which rely on the pressure of an enclosed liquid, are the most common type used in heavy lifting. They come in a variety of configurations, including: bottle jacks, hand- and electric-pump jacks, hydraulic spreaders, and pullers. Inflatable jacks are also used. Other types of jacks that may be used are single-acting low profile (shorty) cylinders with screw and locking features (*Figure 13*), as well as double-acting center hole cylinders for push–pull applications. Cylinders can be combined with pumps (*Figure 14*), hydraulic hoses, and other accessories to form jacking systems such as the ones shown in *Figure 15*.

205F12.EPS

Figure 12 ◆ Hydraulic jacks.

SHORTY CYLINDERS

STANDARD CYLINDERS – 10 TO 500 TON CAPACITY

205F13.EPS

Figure 13 ◆ Hydraulic cylinders.

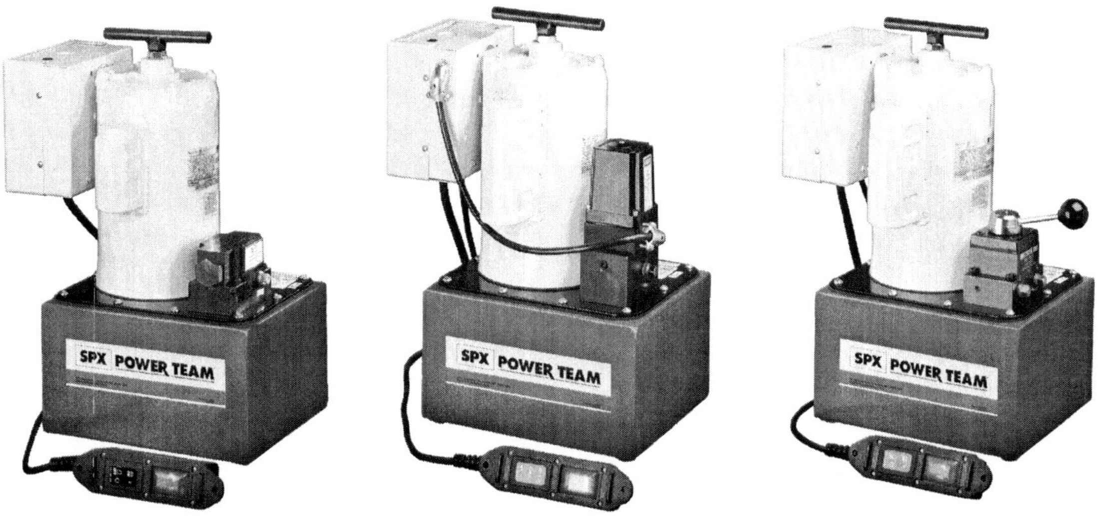

205F14.EPS

Figure 14 ◆ Hydraulic pumps.

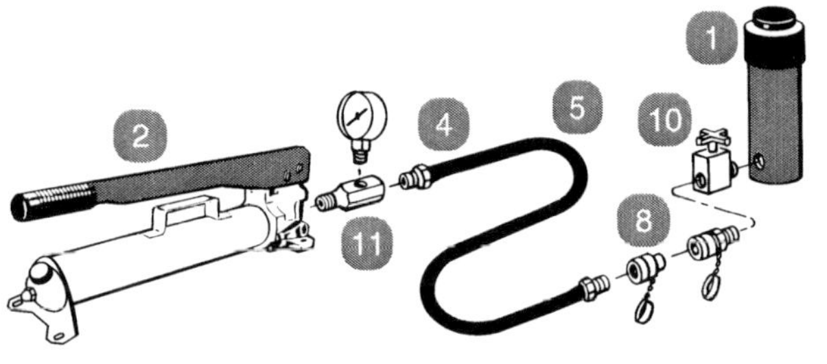

Basic single-acting system with a hand pump, gauge, hose and single-acting cylinder.

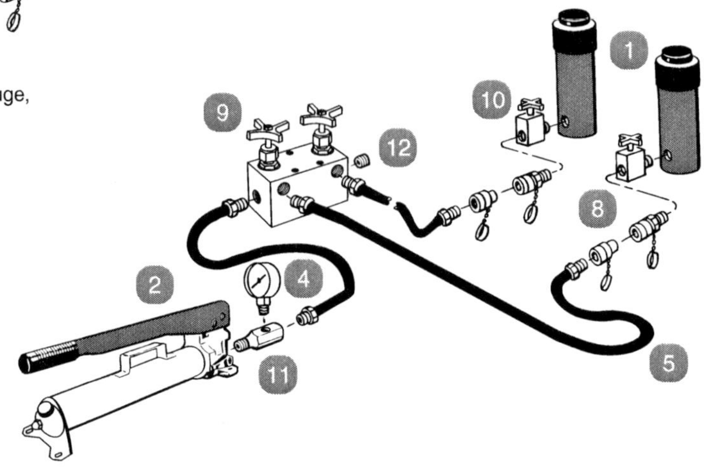

Basic single-acting system with a hand pump, gauge, hose, multiple shut-off valves, load-lowering valves and multiple cylinders.

1. Cylinder – applies hydraulic force.
2. Pump – a device for converting mechanical energy to fluid energy.
3. Directional valve – controls the direction of hydraulic fluid in the system.
4. Gauge – measures P.S.I. pressure (Pounds per Square Inch) and/or force.
5. Hose – transports hydraulic fluid.
6. Manifold – allows distribution of hydraulic fluid from one source to several cylinders.
7. Swivel Connector – allows proper alignment of valves and/or gauges. Used when units being connected cannot be rotated.
8. Quick Coupling – "hose half" and "cylinder half" couplings are used for quick connection and fluid flow check when separated.
9. Shut-Off Valve – regulates the flow of hydraulic fluid to or from cylinders.
10. Load-Lowering Valve– allows metered lowering of cylinder and provides safety when prolonged load holding is required.
11. Tee Gauge Adapter – allows for installation of pressure/tonnage gauge anywhere in the hydraulic system.
12. Pipe Plug – for blocking unused ports within the system.

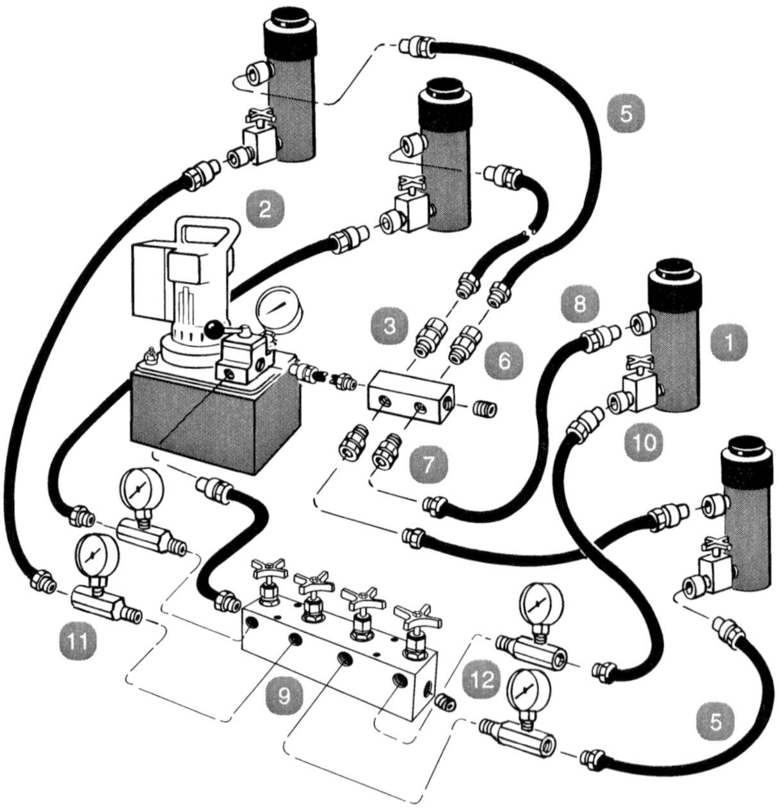

Basic double-acting system with an electric/hydraulic pump, shut-off valves, load-lowering valves and multiple double-acting cylinders.

205F15.EPS

Figure 15 ◆ Jacking systems.

When selecting jacks, the weight, distance, and points of contact of an object are used to determine the type of pumps, cylinders, or other jacking systems to use. A good rule of thumb is to select devices with a capacity about 50 percent greater than the load. This will prevent overloading and damage to the jack, and will also provide a reserve in case the load is underestimated.

When performing jacking operations, make sure the cylinders are fully supported on a firm, non-sliding foundation. Always use hardwood for blocking, and chase the load up or down with the blocking. Plan to have blocking of a thickness that will match the extension of the ram. Make sure there is adequate space for jacks and blocking, and make sure to jack and block at locations that can withstand the load.

While jacking or blocking, stand back occasionally and observe the operation for stability. If it doesn't look right, it probably isn't. Take the necessary steps to make the load stable in order to prevent an accident, even if it means starting over.

3.1.0 Grip Hoists

The grip hoist (*Figure 16*) is a wire rope hoist used for lifting and pulling. In practice, grip hoists in general are not rated or used for vertical lifts. Various models of grip hoists accept wire rope from $\frac{5}{16}$ inch to $\frac{5}{8}$ inch in diameter.

205F16.EPS

Figure 16 ◆ Grip hoist.

The grip hoist is not limited to a certain length of pull or lift, because it continuously accepts cable from an outside source.

Gripping jaws are built into the mechanism to grip the cable as soon as the hand lever is pulled. These jaws hold the load automatically when the pulling is stopped.

The grip hoist has only one hook mounted on the housing. Before operation, this hook must be secured to an object strong enough to support the load being lifted or pulled. The wire rope is fed into the hoist at the opposite end of the housing. Always check the condition of the rope before any lifting or pulling is done.

WARNING!

Failure to secure the hoist to an object strong enough to support the load will cause the load to fall, possibly resulting in injury or death.

3.2.0 Skids

Skids are long timbers placed lengthwise under heavy loads. They perform the following functions:

- Distributing weight over a greater area
- Providing a smooth surface over which equipment can be moved or skidded
- Providing a runway surface for rollers

For most operations, oak timbers 2 or 3 inches thick and about 15 to 20 feet long make good skids. A good practice when using skids is to stagger the joints so that no two sets of planks have their joints in the same place.

The general rule of thumb concerning the line pull needed when using skids on level ground is that approximately 20 percent of the weight of the load should yield the line pull required.

3.3.0 Rollers

The practice of using rollers to move loads is as old as the craft of rigging itself. Hardwood or pipe rollers can be used over skids for some loads (*Figure 17A*). The rollers must be long enough to pass under the load completely. The load must be supported in such a way that the rollers have a smooth surface on which to move. As the load moves along, the free rollers at the rear are brought to the front and placed under the load. To make a turn, the front rollers are moved slightly in the direction of the turn, and the rear rollers are moved slightly in the opposite direction. Mechanical rollers have become much more popular for moving large loads. Jacks are used to raise the load, then the roller assemblies are placed under the load. *Figure 17B* shows examples of mechanical rollers. A huge variety of mechanical rollers are available to move loads ranging into the thousands of tons. A huge variety of mechanical rollers are available to move loads up to the thousands of tons. *Figure 17C* shows an example of rollers in use. Each situation dictates the type to be used.

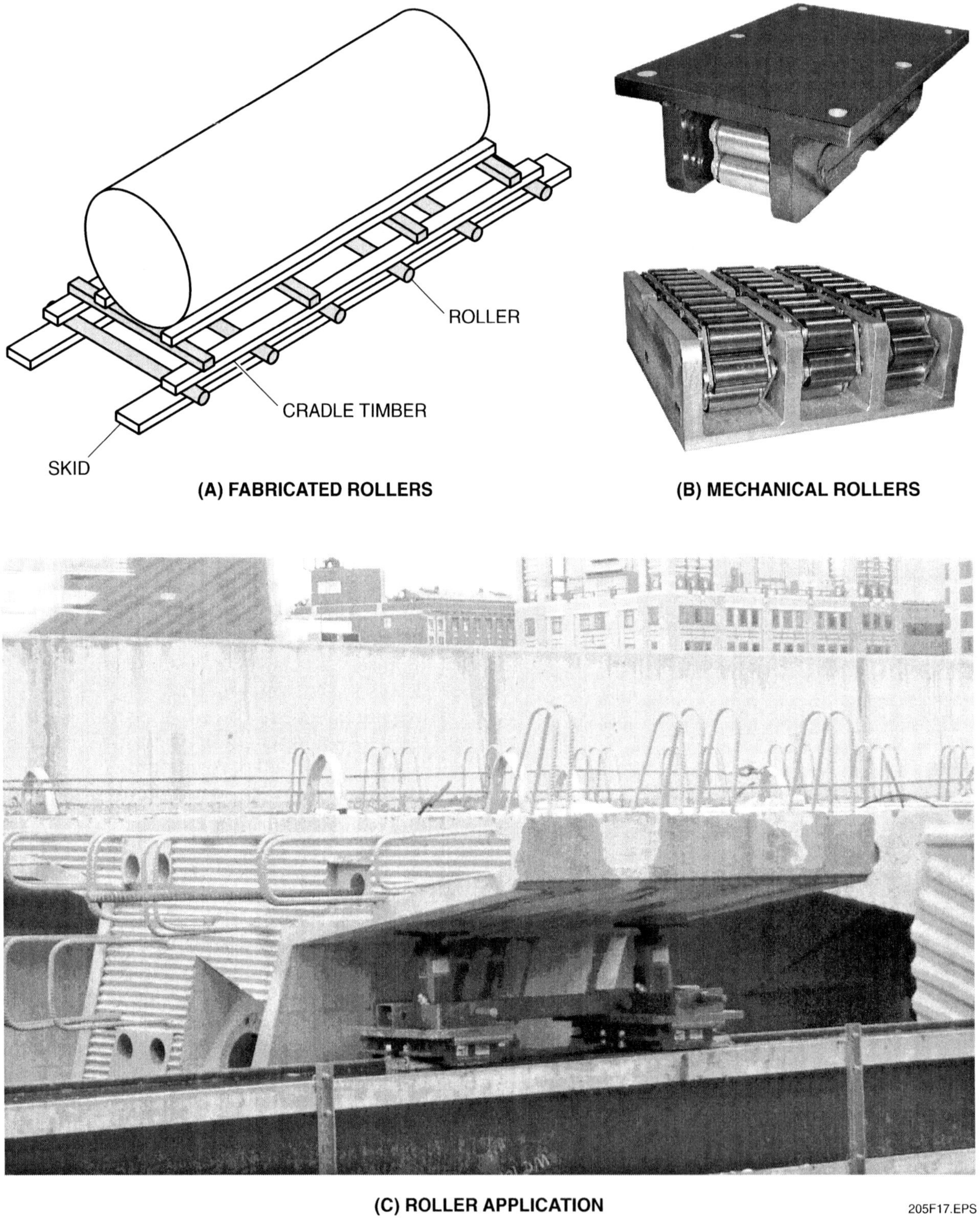

Figure 17 ◆ Use of rollers.

Safety is the most important consideration when moving large loads. The following basic precautions must be observed:

- Do not use metal on metal. Use a softener.
- Avoid pinch points.
- Chock loads properly if they are left on rollers to prevent unintentional movement.
- Use restrainer lines when moving loads on a downward slope.
- Be sure the load is secured to the skid or dolly when transporting it.
- Be certain the dollies and rollers are of sufficient capacity for the load.
- Make sure the surface on which the load is to be moved is as smooth and level as possible.
- Determine the center of gravity to ensure stability before moving a load on rollers or dollies.
- Make sure there are at least three rollers under the load at all times.
- Use a timber track when skidding a load over dirt.
- Apply moving force to the skid, not the load. The force can be applied with a tugger, jack, comealong, or manually with lever bars.

The general rule of thumb concerning the line pull needed when using skids and rollers on level ground is that 10 percent of the weight of the load should yield the required line pull.

3.4.0 Blocking

Blocks should be made of hardwood that is dry and free of defects, oil, and grease. When used with jacks, they should provide a level foundation for the base of the jack. When a jack is being used to lift a metal object, a block should be placed between the object and the head of the jack to prevent slippage. When **chocking** is used to support a circular object, the general rule for safety is that 1 inch of chocking is required for each foot of diameter of the object.

3.5.0 Cribbing

Cribbing is formed by stacking timbers in alternate tiers. It is used to support a heavy weight or to raise an object to a height at which simple blocking would be unstable.

The most important factor to consider when placing cribbing is the foundation. The boom pieces must rest firmly and evenly on the ground. Dig away any high spots rather than filling low areas with loose material.

Cribbing can be used to jack a load in successive stages. The jacks are placed on blocking on the ground. The load is then lifted to the maximum height of the jacks, and one or more layers of cribbing are placed directly under the load. The cribbing is built up until there is enough space between the upper layer of cribbing and the load for an additional layer. At this point, the jacks are lowered until the load rests firmly on the cribbing. The blocking under the jacks is then built up so that the jacks touch the load in their lowered position. The process is then continued as before until the load is jacked to the desired height.

The size of the wood used in both blocking and cribbing usually starts at 4 inches by 4 inches, but may be as large as 14 inches by 14 inches, depending on the application. Hardwood is generally used. The following rules concerning the (safe working load) SWL of wooden beams are necessary information every rigger should know:

- The greater the depth of a timber, the greater the SWL of the beam. Because the SWL varies according to the square of a beam's depth, doubling the depth of a beam increases SWL by four.
- Two similar beams placed together are twice as strong as a single beam. To ensure the strength, secure the ends with nails or dowels.
- The more uniformly distributed a load, the greater the load that can be carried by a beam of a given size, material, and length. A concentrated load applied at the center of a beam reduces the SWL of that beam by one-half.

Figure 18 shows properly stacked cribbing.

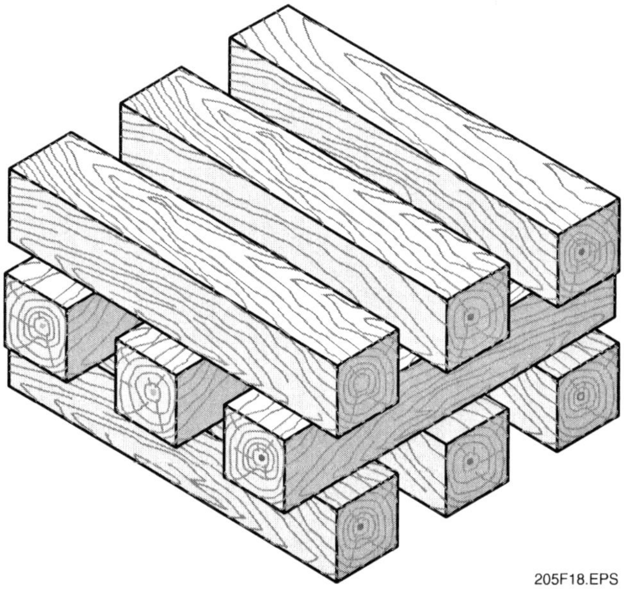

Figure 18 ◆ Cribbing.

3.6.0 Inclined Planes

Inclined planes are also used to raise objects. An inclined plane is a ramp that slopes gradually upward. Less force is required to move an object up an inclined plane than to lift it vertically off the ground. A rigger may use a simple method to figure out the amount of pull required to move an object up an inclined plane if the weight of that object is known. *Figure 19* shows an inclined plane.

Refer to *Figure 19* and follow these steps to calculate the amount of pull required to move an object up an inclined plane:

Step 1 Determine the weight of the object.

Step 2 Draw a sketch of the move, choosing a convenient scale. The sketch will be a right triangle, ABC. In the example shown in Figure 19, the scale used is ¼ inch equals 1 foot.

Step 3 Choose an appropriate scale to represent weight. In the example, the scale is 1 inch equals 10,000 pounds.

Step 4 Starting from line AB, draw a vertical line, DE, equal in length to the weight of the object. The line must be at right angles to the base of the triangle, ABC.

Step 5 Draw a line, DF, at right angles to the hypotenuse (the side opposite the right angle) of the triangle, ABC. The hypotenuse is the line AB.

Step 6 Measure the distance EF.

Step 7 Convert the distance EF to weight, using the scale you have chosen for weight.

Step 8 Increase this amount by 5 percent to allow for the effects of friction. This is the amount of pull required.

In the example, line EF is 3⁄16 inch. Converting this to the weight scale, the pull required before allowing for friction is 1,875 pounds. Take 5 percent of this figure to get 93.75 pounds. Round off to 94 pounds and add to 1,875 pounds to yield 1,969 pounds of pull required to move the object up the inclined plane.

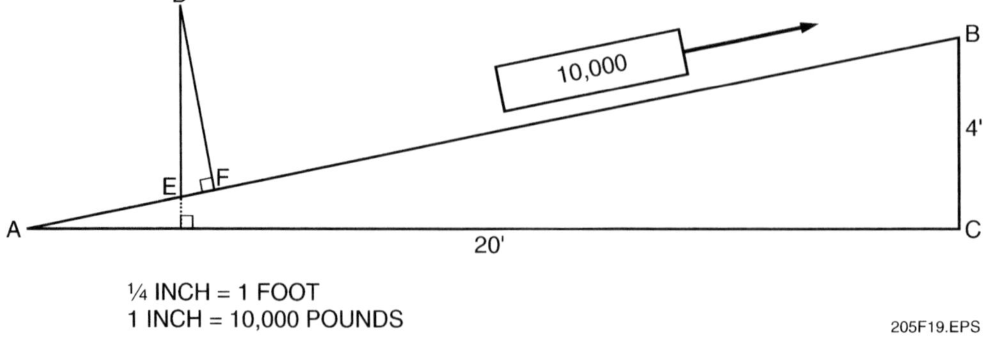

Figure 19 ◆ Inclined plane.

4.0.0 ◆ SLINGS

Some type of sling will be used in nearly every lift you make. It is important that you know how to select the correct sling for each situation, how to properly use the sling, and how to apply derating factors to compensate for sling angles.

Figure 20 illustrates the ways in which slings are measured. Note especially that the overall length of a sling does not take into account the diameters of the wire rope or any special fittings at either end. Note also that the width of the eye of a sling is measured from inside to inside at the widest point and that the width of all fittings is also measured from inside to inside.

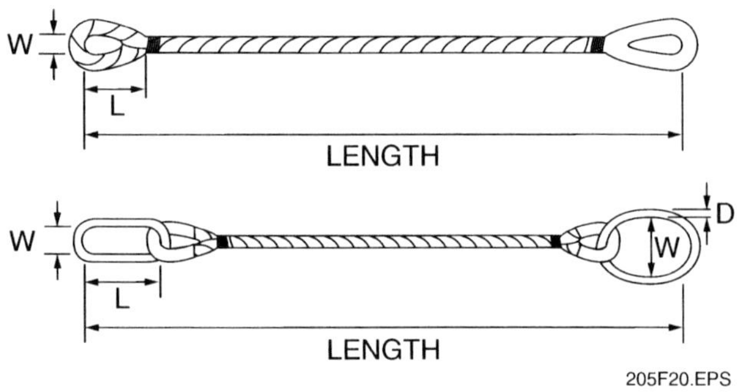

Figure 20 ◆ Sling measurement.

The following are some tips about selecting and using slings:

- Keep written records.
- Have a competent person inspect slings prior to and during their use.
- Use, test, maintain, and inspect all slings according to *29 CFR 1910.184*.

- Remove damaged and defective slings from service.
- Store slings in a cool, dry place out of direct sunlight.
- Do not load slings beyond their rated capacity.
- Do not use slings in temperatures above the manufacturer's recommendations.
- Select the type of sling best suited for the load and conditions.
- Balance and secure loads to prevent slippage when using slings in a basket hitch.
- Make load attachments above the center of gravity or stabilize them.
- Securely attach slings to their loads.
- Protect slings from sharp surfaces.
- Use a shackle when more than one sling is used at the same attachment point.
- Do not drag slings on the floor or other abrasive surfaces.
- Start and stop slowly—shock loading is prohibited.
- Lift the load a few inches, and then check the rigging.
- Do not pull slings from under the load when the load is resting on the sling.
- Select a hitch that will hold and control the load.
- Allow for low D/d ratios.
- Secure all unused sling legs when using more than one on a hook.
- Allow for increased tension caused by sling angles.
- Allow for reductions when using choker hitches.
- Allow for sling angles when forcing (seating) a choker or turning a load.
- Do not use wire rope clips to form eyes in slings.
- Keep fingers and hands clear of loads about to be lifted.
- Wear gloves when handling wire rope slings.

4.1.0 Sling Tensions

The sling angle is the angle formed by the legs of the sling with respect to the horizontal when tension is put upon the load. When using sling angle charts, it is important to know which angle is measured. Because the sling angle directly affects the rated capacity of any sling, proper sling angles are crucial to safe rigging (see *Figure 21*).

A sling with two legs used to lift a 1,000-pound object will carry a 500-pound load in each leg if the sling angle is 90 degrees. As the sling angle changes, the amount of load carried by each leg increases until, at 30 degrees to horizontal (60 degrees to vertical), each leg of the sling is carrying the full weight of the load. Angles less than 30 degrees to horizontal (more than 60 degrees to vertical) will cause the legs of the sling to bear more than the load actually weighs. This will result either in excessive wear of the sling or in total failure of the sling.

ANGLE FACTORS

ANGLE*	FACTOR
90 DEGREES	1.000
80 DEGREES	1.015
70 DEGREES	1.064
60 DEGREES	1.155
50 DEGREES	1.305
40 DEGREES	1.555
30 DEGREES	2.000

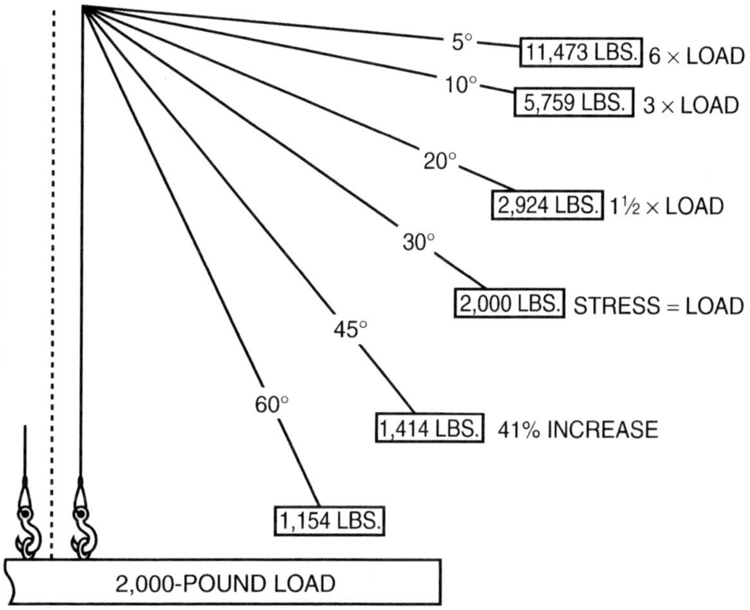

* SLING ANGLES ARE MEASURED FROM HORIZONTAL.

205F21.EPS

Figure 21 ◆ Sling angle factors.

Sling angles should be kept to greater than 45 degrees to horizontal, if possible. Optimum sling angles fall between 60 and 45 degrees to horizontal. Angles beyond this range should be considered hazardous.

Table 1 is an example of a sling capacity chart. This type of chart is an effective tool for determining the safe working load of a sling, depending on its material makeup, construction, size, hitch configuration, quantity, and angle of use.

Table 1 Example of a Sling Capacity Chart

Wire Rope Sling Capacities (LBS) - Flemish Eye - ANSI B30.9

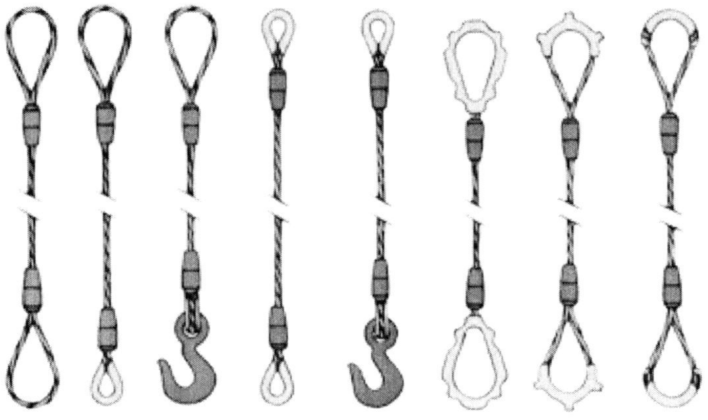

6 x 19 and 6 x 37 Improved Plow Steel IWRC 5/1 Design Factor

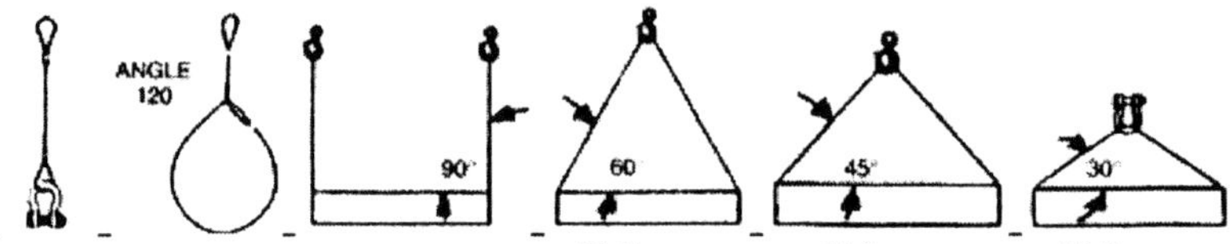

Wire Rope Size	Q&T Carbon Shackle Min. Shackle Size For A D/d >1 Shackle Size	Vertical	Choker	Two Leg	60 Degree Sling Angle	45 Degree Sling Angle	30 Degree Sling Angle
1/4	5/16	1120	820	2200	1940	1500	1100
5/16	3/8	1740	1280	3400	3000	2400	1700
3/8	7/16	2400	1840	4800	4200	3400	2400
7/16	1/2	3400	2400	6800	5800	4800	3400
1/2	5/8	4400	3200	8800	7600	6200	4400
9/16	5/8	5600	4000	11200	9600	7900	5600
5/8	3/4	6800	5000	13600	11800	9600	6800
3/4	7/8	9800	7200	19600	16900	13800	9800
7/8	1	13200	9600	26400	22800	18600	13200
1	1 1/8	17000	12600	34000	30000	24000	17000
1 1/8	1 1/4	20000	15800	40000	34600	28300	20000
1 1/4	1 3/8	26000	19400	52000	45000	36700	26000
1 3/8	1 1/2	30000	24000	60000	52000	42400	30000

* Rated capacities based on pin diameter on larger than the natural eye width or less than the nominal sling diameter.

Refer to ANSI B30.9 For Full Details

Horizontal Sling Angles Of Less Than 30 Degrees Are Not Recommended

205T01.EPS

The exact capacity of the sling can be calculated by determining the sling size, classification, hitch configuration, and sling angle. No rigger is expected to memorize long tables of the safe working loads (SWL) for various configurations of slings. Such information is readily available from a variety of sources. There are, however, rules of thumb that riggers can use to determine the safety of a sling configuration.

> **NOTE**
> Pocket references such as *Bob's Rigging and Crane Handbook* are very useful for quickly calculating sling loads, as well as a wide variety of other rigging data.

Always refer to the manufacturer specifications when calculating SWL. Should the need arise in the field to make a quick determination of SWL, the following formula can be used. First, convert the diameter of the wire rope to eighths; then square the eighths and multiply by 150. The answer is the SWL of the wire rope when choked.

The SWL formulas for slings given here are based on the SWL of the single vertical hitch sling. It is highly recommended that riggers familiarize themselves with this type of configuration. The SWL of the single vertical hitch is based on the diameter of the wire rope and the type of wire rope being used. Every rigger should have an idea of the capacities of different sizes and types of wire rope.

4.2.0 Bridle Hitches

To determine the SWL for bridle hitches, measure the length of the sling legs (L) and the distance between the hook and the load (H). *Figure 22* shows a bridle hitch arrangement.

When this information is obtained, use the use the following formula:

$$SWL = (W)(H \div L)(2)$$

This formula is for symmetrical loads. W is the SWL of one sling in a straight pull. It is strongly recommended that the formula be used for both legs since there is no way of knowing for certain that each leg is supporting an equal portion of the load.

Rigging an asymmetrical load requires a much more precise formula to calculate sling stresses. As shown in *Figure 23*, a 24,000-pound unequal leg angular load is rigged with two slings. The length of sling A is 29.7, and the length of sling B is 12. The distance from the sling A attachment point to directly below the lifting hook, D_1, is 19. The distance from the sling B attachment point to directly below the lifting hook, D_2, is 5, and the height from the sling A attachment point to the lifting hook, H_1, is 22.8. The height from sling B attachment point to the lifting hook, H_2, is 10.9. Entering the numbers into the formula as shown in *Figure 22*, we find that the load stress on sling A is 11,099 pounds, and the load stress on sling B is 17,041 pounds.

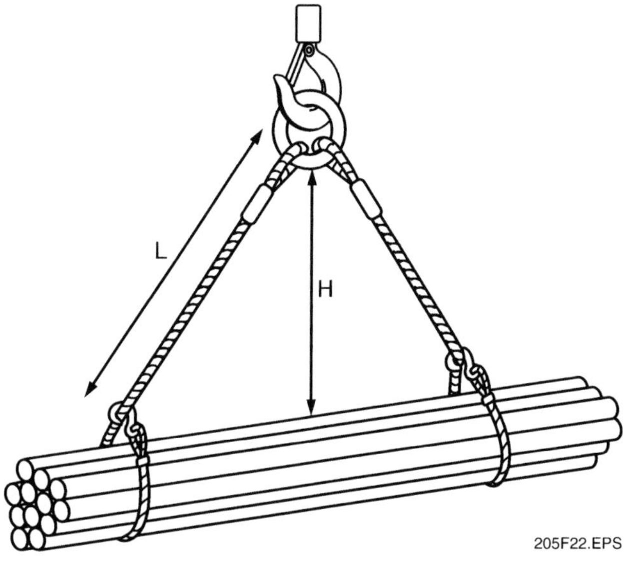

Figure 22 ◆ Bridle hitch arrangement.

4.3.0 Basket Hitches

Basket hitches include single, double, and double-wrapped arrangements. The following guidelines are used to determine the SWL of basket hitches:

- *Single basket hitch* – For vertical legs, take the SWL of the single vertical hitch and multiply it by 2. For inclined legs, the formula is the same as that for the bridle hitch.
- *Double basket hitch* – For vertical legs, multiply the SWL of the single vertical hitch by 4. For inclined legs, the formula is:

$$(SWL)(H \div L)(4)$$

The rigger must verify that all 4 legs carry the load equally. Consider all D/d ratios.

> **NOTE**
> When using this formula, you must make sure that all four legs of the sling carry the load equally. Do not forget the D/d ratio.

- *Double-wrapped basket hitch* – The SWLs for double-wrapped basket hitches are the same as those for the double basket hitches.

ASYMMETRICAL BRIDLE SLING FORMULA

A MUCH MORE ACCURATE FORMULA TO CALCULATE SLING STRESS IS:

$$\text{SLING A} = \frac{\text{LOAD} \times D_2L_1}{D_2H_1 + D_1H_2}$$

$$\text{SLING B} = \frac{\text{LOAD} \times D_1L_2}{D_2H_1 + D_1H_2}$$

$$T_1 = \frac{\text{LOAD} \times D_2L_1}{D_2H_1 + D_1H_2}$$

$$T_2 = \frac{\text{LOAD} \times D_1L_2}{D_2H_1 + D_1H_2}$$

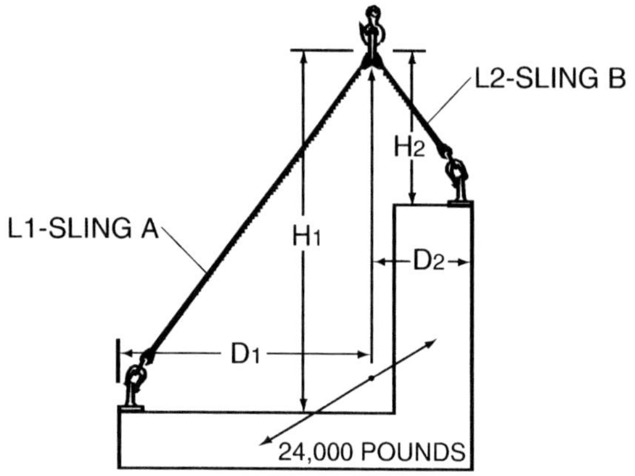

ASYMMETRICAL SLING EXAMPLE

WEIGHT = 24,000 POUNDS

$L_1 = 29.7$ (SLING A) $L_2 = 12$ (SLING B)

$D_1 = 19$ $D_2 = 5$

$H_1 = 22.8$ $H_2 = 10.9$

$$\text{SLING A} = \frac{\text{LOAD} \times D_2L_1}{D_2H_1 + D_1H_2}$$

$$\text{SLING A} = \frac{24{,}000 \times 5 \times 29.7}{(5 \times 22.8) + (19 \times 10.9)}$$

$$\text{SLING A} = \frac{3{,}564{,}000}{321.1}$$

SLING A = 11,099 POUNDS

$$\text{SLING B} = \frac{\text{LOAD} \times D_1L_2}{D_2H_1 + D_1H_2}$$

$$\text{SLING B} = \frac{24{,}000 \times 19 \times 12}{(5 \times 22.8) + (19 \times 10.9)}$$

$$\text{SLING B} = \frac{5{,}472{,}000}{321.1}$$

SLING B = 17,041 POUNDS

NOTE: IRREGULAR LOADS CAN CREATE UNEXPECTED STRESSES. ALWAYS USE A MINIMUM 6:1 SAFETY FACTOR FOR THE RIGGING COMPONENTS.

205F23.EPS

Figure 23 ◆ Asymmetrical bridle sling stress arrangement and formula.

4.4.0 Choker Hitches

Choker hitches include single leg, double leg, and double-wrapped arrangements. The two legs forming the choke naturally tighten to an approximate 45-degree sling angle inside the choke *(Figure 24)*. Always refer to manufacturer specs for the SWL of choker hitches. The following guidelines are used to determine the SWL of choker hitches:

- *Single-leg choker hitch* – For leg angles of 45 degrees or less inside the choke, multiply the SWL of the single vertical hitch by ¾. This is proven by using the formula:

  SWL of a single vertical hitch × (A ÷ B)

  The same formula applies if it is necessary to tighten the choke to the point where the leg angles fall to less than 45 degrees.

  If a double choker hitch is used the sling must be further derated as discussed in the section on bridle hitches.
- *Double-leg choker hitch* – When using more than one sling with a choker hitch, use the single-leg choker hitch formula combined with the calculations for bridle hitches. The following formula applies:

  SWL of a single vertical hitch × (A ÷ B) × (H ÷ L) × 2

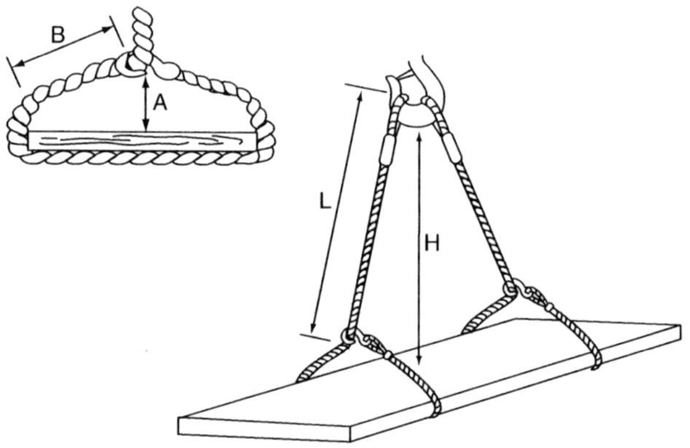

205F24.EPS

Figure 24 ◆ Choker hitch arrangement.

- *Double-wrapped choker hitch* – The values for the double-wrapped choker hitch are the same as those for the single choker hitch or the double choker hitch, depending on the configuration.
- *Rendering the choker hitch* – The capacity of a choker hitch is affected by the angle at the point of choke (*Figure 25*). If the choker is forced down to tighten the eye, it must be derated. This commonly happens when turning loads or erecting oddly shaped loads. The formula is:

  SWL of a single vertical hitch × (A ÷ B) × (H ÷ L) × 2 × percent for angle of choke

Angle Of Choke (in Degrees)	Rated Capacity IWRC & F.C. Rope Percent*
121-135	95
91-121	82
61- 90	71
31- 60	60
0- 30	48

*Percent of sling capacity in choker hitch.

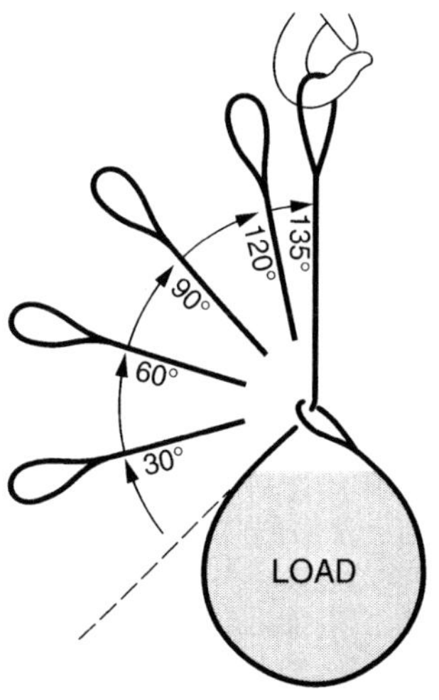

205F25.EPS

Figure 25 ◆ Choker hitch derating factors.

It is easy to overload a choker hitch if the many capacity reducing factors are not considered for every application.

5.0.0 ◆ USING BEAMS

There are many instances when slings alone are not appropriate for a given lift, especially when rigging large, unwieldy loads that might tilt or slip. Beams are used to distribute the weight of a load equally. While the beam configuration will be determined by an engineer in most cases, the rigger must be able to calculate the sling angle appropriate to the load and the length of sling needed to achieve the desired angle.

All lifting beams are required to have their safe working load (SWL) and tare weight stenciled on the beam.

Spreader beams are used to support long or unwieldy loads such as packs of roof sheeting and bundles of rebar. If correctly used, spreader beams help protect the load from tipping, sliding, or bending. Spreader beams also help reduce sling angles, as well as the tendency of slings to crush long loads. *Figure 26* shows a spreader beam being used to protect the heat exchanger tubing bundle while it is being lifted (so that the individual chokers do not damage the tubes). The spreader beam allows the load to be picked up at the ends.

Recall that shoulderless eyebolts are not rated for angular lifting pulls. Loads with such eyebolts must be lifted vertically. The same applies to most welded lifting lugs. Spreader beams are used in such situations. In general, lifting lugs will withstand more force when lifted from directly above, than they will when lifted on an angle.

Spreader beams are normally designed by engineers to meet a particular standard and are test-loaded to a 10 percent overload before they are used to make a lift. Spreader beams used in offshore applications would be tested to a far higher overload, due to the inherent higher dynamic forces resulting from snatching loads in rough sea conditions.

205F26.EPS

Figure 26 ◆ Spreader beam application.

NOTE

ANSI/ASME Standard B30.20 defines the requirements for spreader beams and other below-the-hook devices. Spreader beams must be designed, marked in accordance with the standard, and tested by he manufacturer at 125 percent of their rated capacity.

The use of multiple spreader bars provides better load control and more even distribution of the load for certain types of lifts. *Figure 27* shows a lift using multiple spreader bars.

205F27.EPS

Figure 27 ◆ Multiple spreader beam application.

5.1.0 Adjustable Beams

Adjustable beams have become very popular because they allow a single beam to be used for a variety of lifts. Before the availability of adjustable beams, it was often necessary for riggers to keep a large number of beams of different sizes on hand to deal with the many sizes and types of loads they might encounter. Some adjustable beams, such as the one shown in *Figure 28*, are made of end caps and lengths of pipe. The diameter and length of the pipe determine the lifting capacity. The same caps can be used with different lengths of pipe. Typical pipe diameters used for this purpose are 6 and 8 inches. The beam manufacturer provides a capacity chart that relates length and size of pipe to load capacity. Other types of adjustable beams have a fixed length, but provide several connection points along the length of the beam, as shown in *Figure 28B*. Lifting capacities of adjustable beams range from ½ ton to about 4,000 tons. At the high end of

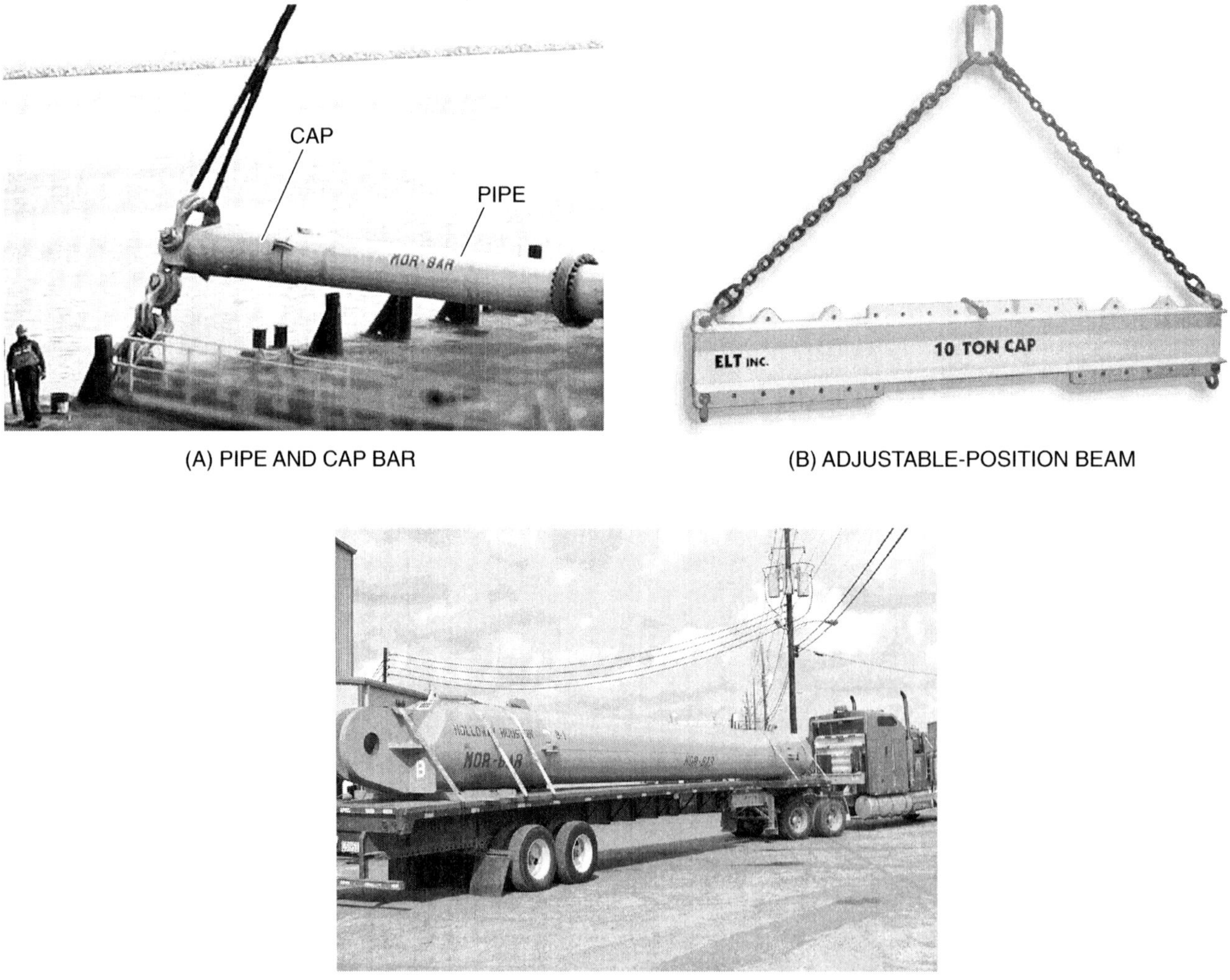

205F28.EPS

Figure 28 ◆ Adjustable lifting beam.

this range, the pipe would be 48 inches in diameter and would be capable of being extended up to 40 feet (*Figure 28C*).

5.2.0 Equalizer Beams

Equalizer beams are often used to distribute the weight of a load between two cranes when making tandem crane lifts. This subject was discussed in detail in the *Lift Planning* module. Equalizer beams are normally designed by an engineer. The safe working load and the tare weight of the beam must be stenciled on the beam. The lift chains or wire rope slings are usually captive to the beam. The beam may be stored on a purpose-built rack when out of use.

6.0.0 ◆ RIGGING REBAR BUNDLES

Reinforcing bars usually arrive at the job site on flatbed trucks. If a spur track is available or adjacent to the area, shipment may be received on railroad cars.

Reinforcement bars or **rebar** are typically delivered to a storage area, where a foreman directs unloading. If the rebar delivery is scheduled to meet the daily placement requirements, the loads are delivered to the points of placement or to the crane so that the bundles can be hoisted to the placement area.

A bill of lading, which accompanies each shipment, lists the materials contained in the shipment. As the material is unloaded, the bundles must be checked against the shipping list to account for all materials. The weight and other necessary information should be recorded for each shipment and the trucks should be unloaded promptly so that they can be released. This prevents additional charges.

6.1.0 Unloading Procedure

Timbers are placed on the ground to support the bundles and keep them free from mud and other job site debris. Blocking and timbers must be selected with care. Hardwood that is large enough to allow the materials to be safely blocked or cribbed should be used. Blocking that has rounded corners or shows signs of dry rot should not be used.

WARNING!
If a load requires more than two people to carry it, lifting equipment must be used to prevent personal injury.

Labor can be saved if the bundles of steel can be hoisted directly from the truck to the area where the bars will be placed. Always make sure that the area where the load will be landed is strong enough to support the weight. Safe load limits should be established before loading. Make sure that the lifting equipment has the lifting capacity and reach to handle the load and that something is available to pad any sharp edges.

6.2.0 Hoisting Equipment

Power equipment is used to raise or lower rebars and to bridge obstacles that prevent the bars from being manually carried. The type of hoisting equipment depends on the size of the job, site congestion, and structure height. The most commonly used hoists are the truck crane, the hydraulic crane, and the Chicago boom. When hoisting bundles of bars 30 feet or longer, always use a spreader beam to prevent bending of the bars. The spreader beam should be at least half the length of the bars. *Figure 29* shows the use of a spreader beam to lift a bundle of rebar.

When bundles are being hoisted in congested or confined areas, always attach a tag line to one or both ends of the load to guide it as it is hoisted. Fiber rope is typically used for this purpose. Safe hoisting requires a skilled operator taking instructions from an authorized signal person using standard hand signals.

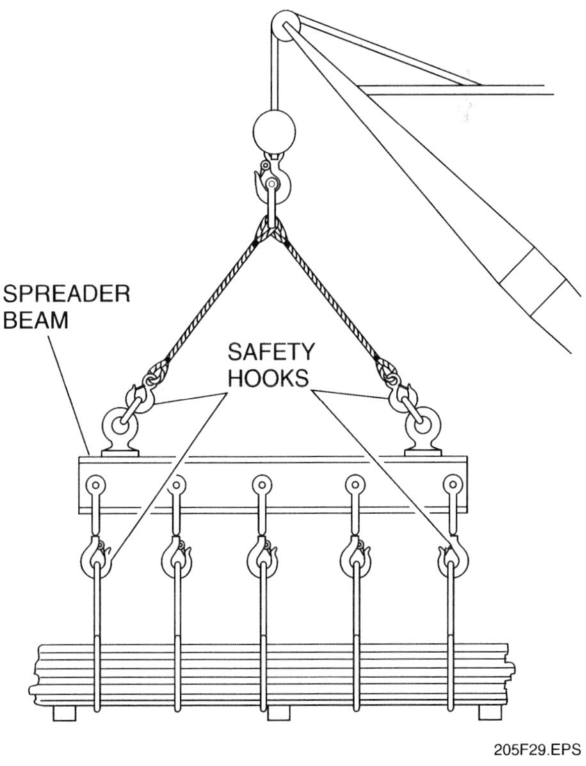

Figure 29 ◆ Use of a spreader beam to rig rebar.

The wire wrapping used to tie the bundle together should never be used to pick up bundles of bars. This wire is not strong enough to carry the load. Use slings made of wire rope at least ½ inch in diameter. The sizes of wire rope hooks and other hardware must be safe for the load capacities.

It is very important that chokers and slings be strong enough to lift the load. The stress on each choker depends on the number of chokers, the angle of the choker, and the total load including the rigging. The total weight to be lifted is divided among the supporting chokers. The closer the angle of the choker to the horizontal, greater the tension in the choker will be. When two slings or chokers are used to lift bundles, thread the loops in the same direction. If one sling is looped from the opposite side, the bundle will twist when hoisted.

Figure 30 shows the proper way to use two slings to lift a bundle.

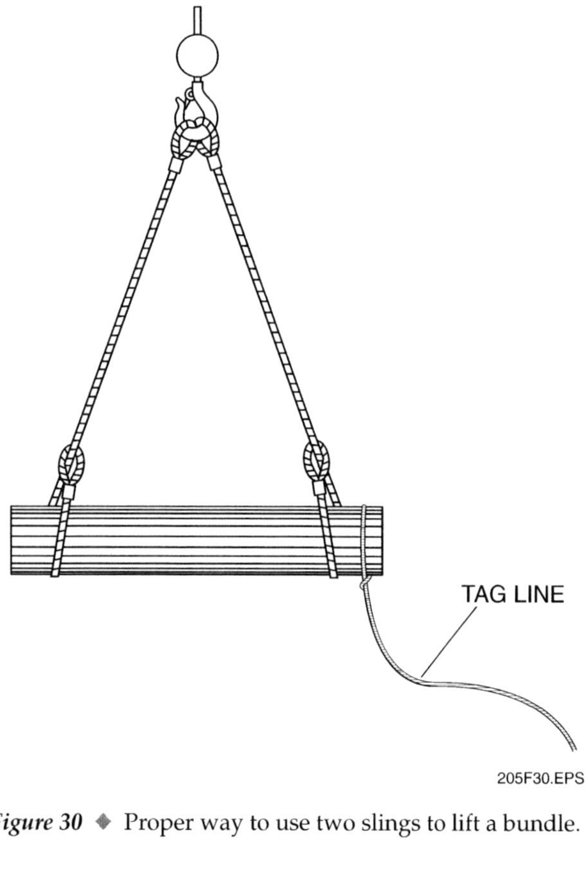

Figure 30 ◆ Proper way to use two slings to lift a bundle.

7.0.0 ◆ SUBMERGED LIFTS

Cranes are often used to lift loads on or near water. This type of lift requires special attention. Submerged lifts involve the load being submerged in a liquid during a lift, which can be hazardous if all load factors are not considered. One factor that must be considered is ground stability. If lifting near water, the ground may be saturated and will therefore be very unstable. Another condition to consider is wave action. As the load comes to the surface during a lift, the movement of the water will make the center of gravity of the load move in relation to the tipping axis. This movement can cause instability of the crane and overload its various components.

Finally, **buoyancy** is an important factor to consider. Buoyancy is the tendency of an object to remain afloat, or the upward pressure of a liquid or gas. The relationship of the weight of an object to the amount of fluid it displaces determines the significance of this effect. If an object displaces an amount of water greater than its weight, then the object (such as an ocean liner) will float. If the object displaces an amount of water less than its weight, then the object (a rock, for example) will sink. If the amount of water displaced is equal to the weight of the object, it will be neutrally buoyant. This is most critical when the load is brought to the surface of the liquid. As the load is brought to the surface, it will lose the effect of buoyancy, and the full weight of the load will be applied to the crane. If this change in weight is not accounted for in the lift calculation, the crane can be significantly overloaded, resulting in structural failure or tipping.

A side effect of submerged lifts, as well as lifts from wet areas, is suction. When a load is lifted from a wet area, the load may become stuck. The load can trap a pocket of air under it, and as the load is lifted, the suction can add unexpected weight to the crane. Another hazard with suction occurs when the load breaks free of the suction. This unexpected loss of load deflects the boom backwards, possibly resulting in the tipping of the crane.

8.0.0 ◆ COMMUNICATION

In *Rigging Fundamentals*, you learned the hand signals used for communication between the rigger and the crane operator. In this module, we will discuss the standard requirements that govern the use of hand, voice, and audible signals, as dictated by *ASME Standard B30.5-2004*.

> **NOTE**
>
> This section summarizes the content of *ASME B30.5-2004*. It is strongly recommended that you read the complete content of this standard as it relates to signaling.

Regardless of the type of signaling being used, the signal person must be tested and must demonstrate their qualifications in the following areas:

- Understanding of crane operations and limitations
- Standard hand signals
- Standard voice signals

Hand signals must be posted on site. If voice signals are being used via electronic communication devices, the following conditions must be met:

- The voice signals must be discussed and agreed upon between the operator and the signaler.
- Electronic communication devices must be tested before use, and spare batteries must be available for battery-powered devices.
- The operator and signal person must contact and identify each other before the lift begins.
- Directions must be given from the operator's perspective; e.g., swing right.
- Each series of signals must contain three elements in the following order:

  1. Function and direction
  2. Distance and/or speed
  3. Function stop

  For example, load down 10 feet, slow, slow, load stop
- The person directing the lift will determine if multiple simultaneous function signals may be used; for example, raising and swinging the load at the same time.

If any special signals or modifications of standard signals need to be used, they must be agreed upon in advance by the lift director, signal person, and crane operator.

An audible signal such as a horn may be used as follows when the machine is being moved:

- *One short signal* – STOP
- *Two short signals* – GO AHEAD
- *Three short signals* – BACK UP

Emergency audible signals using a short or continuous tone may be given by anyone on the site. It is necessary that this signal be standardized for the site.

Summary

Chemical plants, power plants, paper mills, oil refineries, and other industrial operations require the hoisting or movement of immense equipment, machines, and structures. Many such operations require more than one crane. In many instances, it is necessary to move piece of equipment or a structure horizontally, in order to reposition it or to get it into position to be hoisted by a crane. A variety of machines and devices are used for these purposes. The rigging specialist needs to know which type of device to use and must be able to select the device with the correct capacity for the application.

Specialized knowledge is needed to plan and implement complex rigging operations. Many factors can affect a lifting operation, and each factor can cause a load to act in a different manner. It is up to the crane operator and rigging specialist to understand how to prevent conditions that might cause a crane to tip. For example, it is important to know how the tipping axis can change simply by rotating the crane to an over-the-side quadrant. All such factors must be considered during development of the lift plan and during the rigging operation itself.

Although engineers may design the lift, the rigging specialist plays an important role in the planning and in carrying out the lift. At every step in the process, the rigging specialist must ensure that the operation is carried out safely and that the correct equipment and methods are being used.

Review Questions

1. Backward stability of a crane is affected by _____.
   a. high boom angles and inclines
   b. excessive counterweights
   c. over-the-side position
   d. low boom angles and inclines

2. What effect occurs when a load is hoisted from a point other than directly under the load?
   a. Pendulum
   b. Buoyancy
   c. Acceleration
   d. Hydraulic load

3. The sudden release of a load that causes a rapid change in effective load weight is called _____.
   a. wave action
   b. recoil
   c. impact loading
   d. acceleration

4. Doubling the depth of a cribbing beam increases the SWL by a factor of _____.
   a. two
   b. three
   c. four
   d. six

5. A sling angle is considered hazardous if it is less than _____ degrees from horizontal.
   a. 30
   b. 45
   c. 60
   d. 75

6. The formula for determining the SWL of a bridle hitch is _____.
   a. SWL = (W)(L÷H)(2)
   b. SWL = (W)(H÷L)(2)
   c. SWL = (L)(W÷H)(2)
   d. SWL = (H)(L÷W)(2)

7. Adjustable lifting beams are limited to loads of less than 100 tons.
   a. True
   b. False

8. When an audible signal such as a horn is being used for communication, the signal to stop is _____.
   a. one long blast
   b. two short blasts
   c. two long blasts
   d. one short blast

9. Bundles of rebar should be lifted by _____.
   a. the tie wire around the bundle
   b. magnets
   c. tag lines
   d. wire rope slings

10. When a bundle of rebar is being lifted with two chokers, the choker loops should be threaded in the opposite direction to improve load stability.
    a. True
    b. False

GLOSSARY

Trade Terms Introduced in This Module

Acceleration: The rate of change of velocity with respect to time.

Buoyancy: The tendency of an object to remain afloat, or the upward pressure of a liquid or gas.

Center of gravity: The point at which an object is perpendicular to and balanced in relation to the earth's gravitational field.

Chocking: A block or wedge placed under a round object, such as a wheel, to keep it from moving.

Lever: A device that pivots on a fulcrum to move a load.

Leverage: The power applied to a lever to cause a load to move.

Pendulum effect: A condition that occurs when a load is hoisted from a point other than directly under the load line, thus causing the load and hoist line to swing like a pendulum until it finds its center of gravity.

Radius: The horizontal distance from a projection of the axis of rotation to the supporting surface before loading to the center of the vertical hoist line or tackle with the load applied.

Rebar: Abbreviation for reinforcing bar. They are bars generally made of used that placed in concrete forms before the concrete is poured in order to increase the strength of the concrete.

Tipping axis: The fulcrum point on which a lever is balanced.

APPENDIX A

Installation of a Tube Bundle in a Heat Exchanger

Contributed by Cianbro Construction

This describes a project in which a heat exchanger tube bundle was installed in a vertical heat exchanger. The step numbers relate to the photos in *Figures A1* and *A2*.

Step 1 The tube bundle was delivered to the site. The tube bundles was delivered in a strong-back frame to prevent bending, twisting, and other damage.

Step 2 A vertical up-righting strong-back frame was used to raise the tubing bundle. This photo shows the lifting beams that will be used at the top of the frame. Note the size in relation to the worker in the background.

Step 3 The tube bundle was lifted and rotated for placement in the up-righting frame.

Step 4 The tube bundle was placed into the up-righting frame.

Step 5 The tube bundle was pinned into the up-righting frame and the lift begins.

Step 6 The lifting process required two cranes.

Step 7 Once the assembly was vertical, an assist crane was hooked up to separate the uprighting frame from the tube bundle.

Step 8 The up-righting frame and tube bundle were separated.

Step 9 The tube bundle was positioned over the module and lowering began.

Step 10 The tube bundle was placed into the module.

Step 11 A steam drum was lowered onto the tube module section.

STEP 1

STEP 2

STEP 3

STEP 4

STEP 5

STEP 6

205A01.EPS

Figure A-1

STEP 7

STEP 10

STEP 8

STEP 11

STEP 9

205A02.EPS

Figure A-2

APPENDIX B

Removing a Horizontal Heat Exchanger Using an Extractor

Contributed by Exxon Mobil Corporation

In this example, workers are removing a tube bundle from a horizontal heat exchanger working in tight quarters. In a situation like this, a special extractor is used to remove the tube bundle. Before work was begun, rigging specialists visited the job site to plan the removal. During this visit, the rigging team completed the following:

- Verified that there was enough space to do the removal
- Identified obstructions that might hinder the sling length or crane boom movement
- Identified components that would need to be dismantled to make room for the removal
- Identified scaffolding needs
- Identified insulation that would need to be removed

The next step in the process is to do the preparatory work on the site. This includes:

- Obtaining the weights of the exchanger and its components
- Inspecting slings and rigging
- Having scaffolds built and removing insulation as necessary to allow access for bolts and rigging
- After the crane was set up, removing piping and insulation as necessary, and securing other piping using a chain hoist
- Removing the channel box from the end of the exchanger using a ¼-inch × 20-foot flat braided wire rope sling choked around the entire channel box
- Verifying that all lifting lugs were adequately designed and in good condition, then removing the shell cover and floating head using two 3-ton chain falls and beam clamps (Engineering approved all overhead attachment points)

Once the preparations are complete, the extraction process begins:

Step 1 The extractor is rigged to the crane and raised into position to pull the exchanger (*Figure B1*).

Step 2 The extractor is attached to the front of the exchanger shell using two 1-inch eyebolts, two 1-inch shackles, and two ¾-inch × 6-foot rope slings (*Figure B2*).

FIGURE B1

FIGURE B2

205A03A.EPS

Step 3 The extractor has three main components (*Figure B3*). The bail can be moved along the rails to balance the load (see *Figure B2* for comparison). The car is used to pull the exchanger along the rails.

Step 4 The extractor operator uses remote controls to position the bail to keep the crane boom over the load CG (*Figure B4*).

Step 5 Once the tube sheet is completely supported by, and locked onto the car, the extractor is disconnected from the shell, and the load is ready to be lowered to the ground.

FIGURE B3

FIGURE B4

205A03B.EPS

APPENDIX C

Vertical Vessel Rigging

Contributed by Fluor Corporation

A common occurrence in petrochemical and refinery work is the up-ending and setting of vertical vessels. This is commonly done with two cranes: a crane that raises the top end of the vessel that is called the main or lift crane, and a crane that supports the bottom of the vessel while the top is being lifted (*Figure C1*). This one is called the tailing crane. Because of the use of two cranes and the rotation of the vessel 90 degrees during the lift, there are special considerations for the rigger in the selection of the rigging and the performance of the lifts. An example of a typical vertical vessel would be the 40,000-lb vessel shown in *Figure C2*.

The initial pick load (IP) at the top end of the vessel will be:

$IP = (40{,}000) \times (17/32) = 21{,}250$ lbs

The tail load (TL) at the bottom end of the vessel will be:

$TL = (40{,}000) \times (15/32) = 18{,}750$ lbs

Top Rigging Hookup

Normally, top head lugs will be provided for the vessel and the rigger must select shackles, a spreader, and slings, as shown in *Figure C3*. Note that the capacities are determined when the vessel is in the vertical position, and all the weight is supported by the top rigging.

Shackles must be sized for both capacity and to fit the size of the hole as closely as possible. Slings must be selected for capacity, to match the shackle (wire rope should not be bent around a shackle smaller than its own diameter), and appropriate length. The length must provide the correct sling angle (60 degrees in the case shown), and the proper clearances.

The spreader must be the correct length and capacity. It must be designed for the purpose and have the rated capacity permanently attached.

If trunnions are provided instead of lugs, either the sling eye is placed around the trunnion, in which case the trunnion must not be larger than the natural width of the eye, or the sling is placed in a basket hitch around the trunnion. In that case, the trunnion must not be so small that the D/d ratio becomes critical.

Tailing Hookup

If a tailing lug is provided, the rigger simply selects a shackle to meet the capacity and the lug hole diameter, and then selects a sling for capacity and to match the shackle as was done for the top head hookup (see *Figure C4*).

If there is no tailing lug, a choker would be placed around the bottom of the vessel. Wood timbers should be used under the choker as softeners and to help keep the sling from sliding up.

An important point the rigger should be aware of during the up-ending of the vessel is that the tailing load does not drop off until the vessel is near the vertical position (*Figure C5*).

As the chart in *Figure C6* demonstrates, the tailing load decreases only gradually until the near end of the lift.

The selection of the main lift crane will usually require a layout similar to the one in *Figure C7* showing the clearances between the boom and crane and the rigging.

FIGURE C1 – Example of vertical vessel rigging

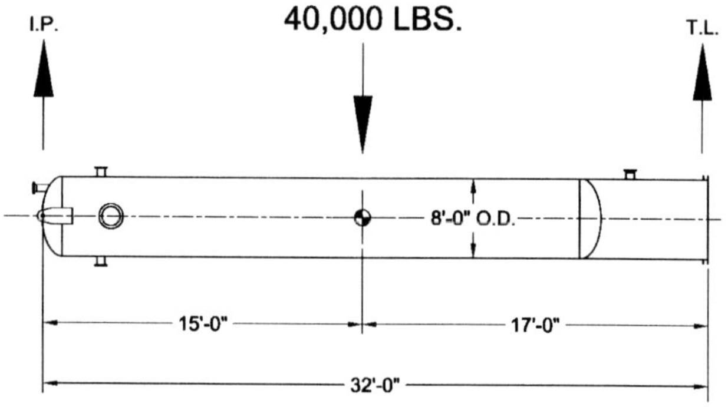

The initial pick load (IP) at the top end of the vessel will be:
IP = (40,000) × (17/32) = 21,250 lbs

The tail load (TL) at the bottom end of the vessel will be:
TL = (40,000) × (15/32) = 18,750 lbs

FIGURE C2 – Vessel Dimensions

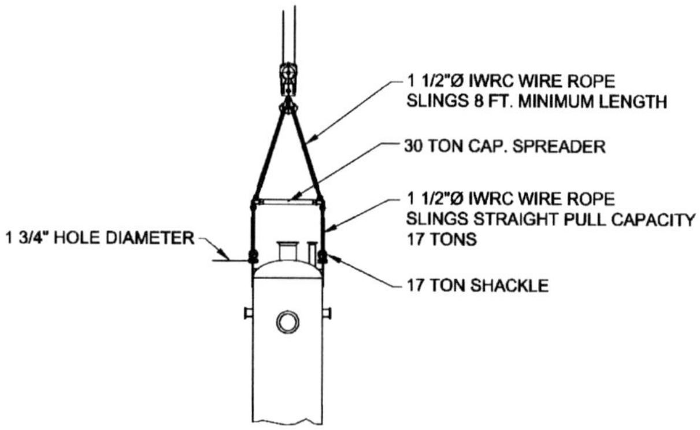

FIGURE C3 – Lifting Hardware

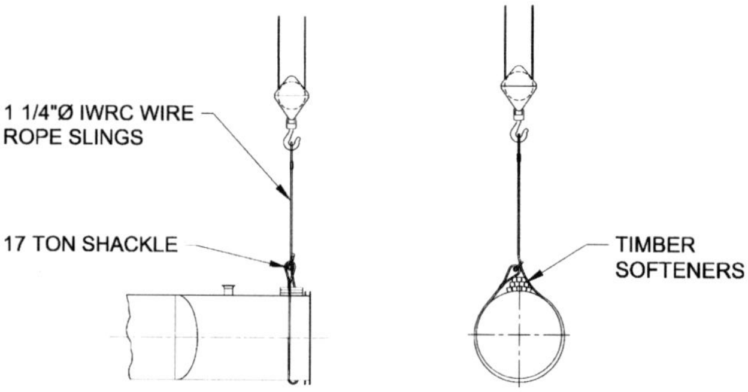

FIGURE C4 – Tailing hookup

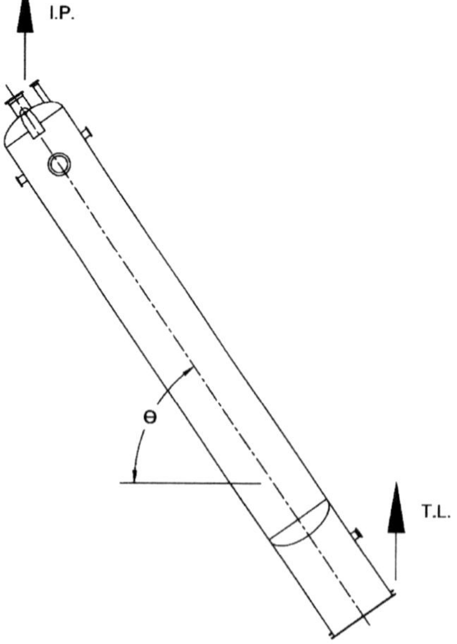

FIGURE C5 – Tailing load

205A04.EPS

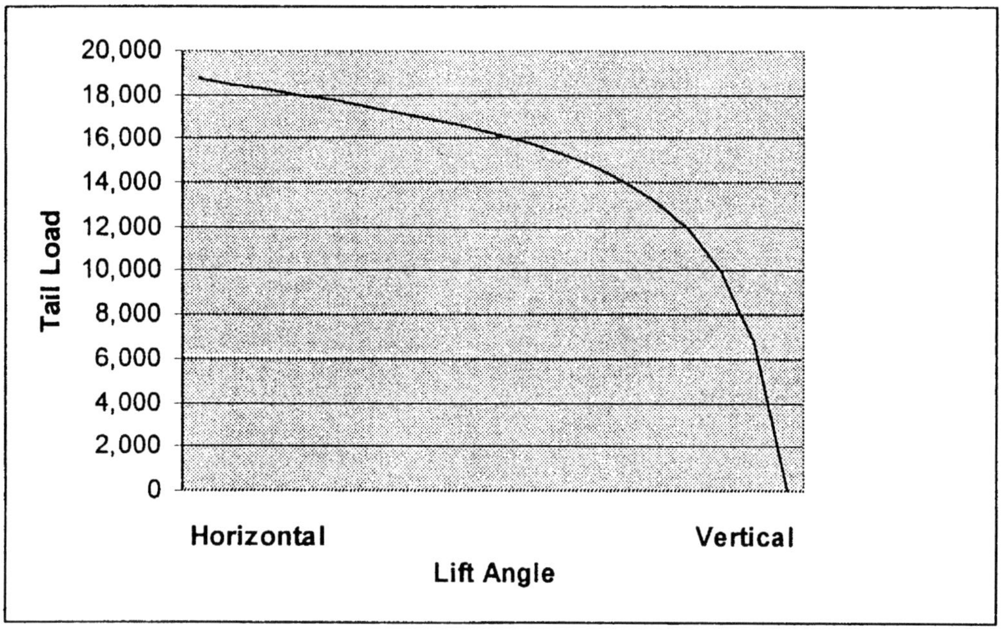

FIGURE C6 – Tailing load chart

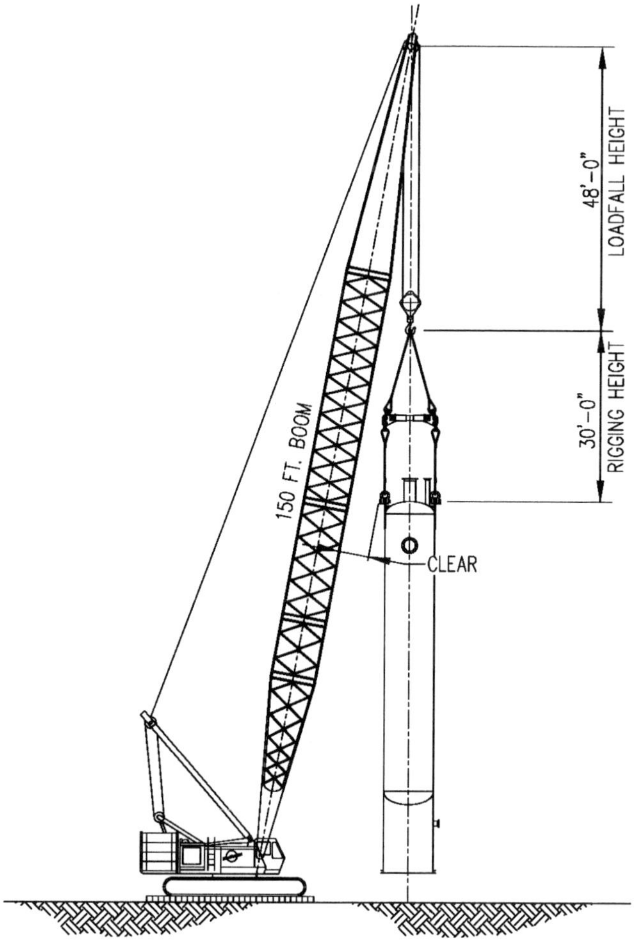

FIGURE C7 – Main lift crane layout

205A05.EPS

Additional Resources

This module is intended to be a thorough resource for task training. The following reference works are suggested for further study. These are optional materials for continued education rather than for task training.

Bob's Rigging and Crane Handbook, Latest Edition. Leawood KS: Pellow Engineering Services.

IPT's Crane and Rigging Handbook, Latest Edition. Edmonton, Alberta, Canada: IPT Publishing and Training.

Figure Credits

Exxon Mobil Corporation	205F01, Appendix B
Holloway Houston	205F02, 205F26, 205F27, 205F28A, C
Lampson International LLC	205F03
Manitowoc Crane Group	205F11
SPX Power Team	205F12–205F15
TorqHoist, Inc.	205F16
Hilman Rollers, Inc.	205F17B, C
Commercial Group Lifting Products	205T01
Engineered Lifting Technologies, www.eltlift.com	205F28B
Cianbro Corporation	Appendix A
Fluor Corporation	Appendix C

CONTREN® LEARNING SERIES — USER UPDATE

The NCCER makes every effort to keep these textbooks up-to-date and free of technical errors. We appreciate your help in this process. If you have an idea for improving this textbook, or if you find an error, a typographical mistake, or an inaccuracy in NCCER's Contren® textbooks, please write us, using this form or a photocopy. Be sure to include the exact module number, page number, a detailed description, and the correction, if applicable. Your input will be brought to the attention of the Technical Review Committee. Thank you for your assistance.

Instructors – If you found that additional materials were necessary in order to teach this module effectively, please let us know so that we may include them in the Equipment/Materials list in the Annotated Instructor's Guide.

Write: Product Development and Revision
National Center for Construction Education and Research
P.O. Box 141104, Gainesville, FL 32614-1104

Fax: 352-334-0932

E-mail: curriculum@nccer.org

Craft ______ Module Name ______

Copyright Date ______ Module Number ______ Page Number(s) ______

Description ______

______

______

______

(Optional) Correction ______

______

______

(Optional) Your Name and Address ______

______

______

Module 38206-06

Personnel Lifts

COURSE MAP

This course map shows all of the modules in the second part of the *Rigging* curriculum, *Rigging*. The suggested training order begins at the bottom and proceeds up. Skill levels increase as you advance on the course map. The local Training Program Sponsor may adjust the training order.

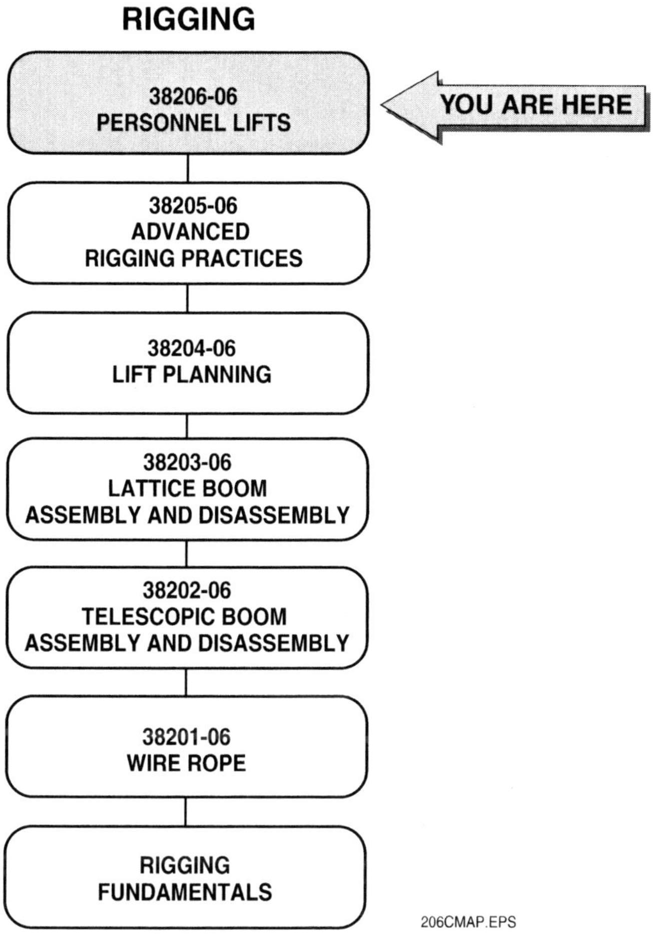

206CMAP.EPS

MODULE 38206-06 CONTENTS

1.0.0 INTRODUCTION 6.1

2.0.0 FALL PROTECTION 6.2

3.0.0 PLATFORM REQUIREMENTS 6.2

3.1.0 Design Criteria 6.2

3.2.0 Platform Specifications 6.2

3.3.0 Personnel Platform Loading 6.3

3.4.0 Rigging 6.3

4.0.0 BOOM-ATTACHED PLATFORM REQUIREMENTS 6.4

5.0.0 CRANE AND OPERATIONAL REQUIREMENTS 6.5

5.1.0 Operational Criteria 6.5

5.2.0 Instruments and Components 6.6

5.3.0 Work Practices 6.6

5.3.1 Working Over or Near Water – OSHA 1926.106 . . . . 6.6

5.4.0 Traveling 6.7

6.0.0 PERSONNEL PLATFORM INSPECTION 6.7

6.1.0 ASME-Prescribed Personnel Platform Inspection 6.7

7.0.0 TRIAL LIFT 6.7

7.1.0 Proof Testing 6.8

8.0.0 ADVANCED OPERATIONS TECHNIQUES FOR HOISTING PERSONNEL 6.8

SUMMARY 6.9

REVIEW QUESTIONS 6.10

GLOSSARY 6.13

APPENDIX A, Personnel Platform Lift Planning Authorization Form . . 6.15

APPENDIX B, Personnel Platform Pre-Lift Inspection 6.16

APPENDIX C, Power Line Awareness Permit 6.17

ACKNOWLEDGMENTS 6.19

Figures

Figure 1 Suspended personnel platform 6.3
Figure 2 Bridle sling using a master link 6.4
Figure 3 Bridle sling using a shackle .6.4
Figure 4 Prohibited zone and avoidance zone 6.9
Figure 5 Moving a crane through an avoidance zone 6.9

Tables

Table 1 High-Voltage Power Line Clearances 6.8

MODULE 38206-06

Personnel Lifts

Objectives

When you have completed this module, you will be able to do the following:

1. Identify which federal regulations apply to hoisting personnel.
2. Identify which consensus standards apply to hoisting personnel.
3. Visually inspect the platform, suspension system, and attachment points.
4. Define operation techniques for hoisting personnel near power lines.

Prerequisites

Before you begin this module, it is recommended that you successfully complete *Rigging Fundamentals* and *Rigging,* Modules 38201-06 through 38205-06.

Required Trainee Materials

1. Pencil and paper
2. Appropriate personal protective equipment
3. Copies of OSHA and ASME regulations

1.0.0 ◆ INTRODUCTION

Like every other lift operation, hoisting personnel requires careful attention to detail and safety. This lifting practice and aspects related to it are covered under *OSHA 29 CFR 1926.104, 1926.106, 1926.550(g), 1926.753,* and *ASME B30.23.* The purpose of this module is to cover in-depth the guidelines of both the federal regulations and the consensus standard.

The intent of *29 CFR 1926.550* is to establish minimum design criteria, equipment features, and operational procedures when using material hoisting equipment to lift personnel.

Although using cranes to lift people was common in the past, OSHA regulations, as spelled out in *29 CFR 1926.550(g)(2),* now prohibit the practice unless no other method is suitable. However, for steel erection, *29 CFR 1926.753(c)(4)* specifically exempts the requirement of *29 CFR 1926.550(g)(2)* and allows lifting of personnel. When lifting is permitted, certain controls must be in place, including the following:

- Depending on the type of rope, the load rope must be capable of supporting at least seven or ten times the maximum intended load.
- Cranes must be equipped with a boom angle indicator and, if applicable, a boom length indicator.
- The hoist drum must have a controlled load-lowering system or device other than the load hoist brake. Free fall is prohibited.
- Crane brakes and locking devices such as pawls or dogs must be engaged when the occupied suspended personnel platform is in a stationary position.
- The crane must be level within one percent of level grade and located on firm footing. Any outriggers must be fully deployed.
- The crane capacity is cut in half for the operating radius used.
- Anti-two-blocking devices are required.
- The suspended personnel platform must be specifically designed for lifting personnel.
- Before the lift is used, it must be tested with a comparable weight and then inspected.
- Every intended lift must undergo a trial run with weights rather than people.

2.0.0 ◆ FALL PROTECTION

Fall protection equipment and guidelines for hoisting personnel are covered under *OSHA 29 CFR 1926.104, Full-Body Harnesses, Lifelines, and Lanyards*. These guidelines are as follows:

- Lifelines, full-body harnesses, and **lanyards** shall be used only for employee safeguarding. Any lifeline, full-body harness, or lanyard actually subjected to in-service loading, as distinguished from static load testing, shall be immediately removed from service and shall not be used again for employee safeguarding.
- Lifelines shall be secured above the point of operation to an **anchorage** or structural member capable of supporting a minimum dead weight of 5,400 pounds.
- Lifelines used on rock-scaling operations, or in areas where the lifeline may be subjected to cutting or abrasion, shall be a minimum of ⅞-inch wire core manila rope. For all other lifeline applications, a minimum of ¾-inch manila or equivalent with a minimum breaking strength of 5,400 pounds shall be used.
- Full-body harness lanyards shall be a minimum of ½-inch nylon or equivalent, with a maximum length to provide for a fall of no greater than six feet. The lanyard shall have a nominal breaking strength of 5,400 pounds.
- All full-body harness and lanyard hardware shall be drop-forged or pressed steel, cadmium-plated in accordance with Type 1, Class B plating specified in Federal Specification *QQ-P-416*. The surface shall be smooth and free of sharp edges.
- All full-body harness and lanyard hardware, except rivets, shall be capable of withstanding a tensile loading of 4,000 pounds without cracking, breaking, or taking a permanent deformation.

Employees who may be exposed to fall hazards must be trained to recognize the hazards to which they may be exposed. They must also be able to demonstrate proficiency in the use of fall protection systems. The following are among the training requirements:

- Understanding the general requirements of the OSHA fall protection standard
- Knowing the nature of fall hazards in the work area
- Being able to perform the correct procedures for erecting, maintaining, disassembling, inspecting, and using fall protection systems

When applicable, training must also be provided in the following areas:

- The role of each employee in the safety monitoring system
- Limitations on the use of mechanical equipment during the performance of elevated work
- Correct procedures for the handling and storage of equipment and materials
- The role of employees in fall protection plans

3.0.0 ◆ PLATFORM REQUIREMENTS

In compliance with *OSHA 29 CFR 1926.550*, platform requirements are broken down into four categories: design criteria [*29 CFR 1926.550(g)(4)(i)*], platform specifications [*29 CFR 1926.550(g)(4)(ii)*], platform loading [*29 CFR 1926.550(g)(4)(iii)*], and rigging [*29 CFR 1926.550(g)(4)(iv)*]. These requirements are detailed in the following sections.

3.1.0 Design Criteria

The suspended personnel platform, also called a manbasket or workbasket (*Figure 1*), shall be designed by a qualified engineer or a competent person qualified in structural design.

The suspension system shall be designed to minimize tipping of the platform due to movement of employees occupying the platform.

The personnel platform itself, except the **guardrail** system and personal fall-arrest system anchorages, shall be capable of supporting, without failure, its own weight and at least five times the maximum intended load.

NOTE

In the following sections taken from OSHA and ASME standards, a suspended personnel platform may be called a platform or a personnel platform.

3.2.0 Platform Specifications

Each personnel platform shall be equipped with a guardrail system and shall be enclosed at least from the toeboard to mid-rail with either solid construction or expanded metal having openings no greater than ½ inch (1.27 cm). Other platform specifications are as follows:

- A grab rail shall be installed inside the entire perimeter of the personnel platform.
- Access gates, if installed, shall not swing outward during hoisting.
- Access gates, including sliding or folding gates, shall be equipped with a restraining device to prevent accidental opening.

Figure 1 ◆ Suspended personnel platform.

- Head room shall be provided which allows employees to stand upright in the platform.
- In addition to the use of hard hats, employees shall be guarded by overhead protection on the personnel platform when employees are exposed to falling objects.
- All rough edges exposed to contact by employees shall be surfaced or smoothed in order to prevent injury to employees from punctures or lacerations.
- All welding of the personnel platform and its components shall be performed by a qualified welder familiar with the weld grades, types, and material specified in the platform design.
- The personnel platform shall be conspicuously posted with a plate or other permanent marking which indicates the weight of the platform and its rated load capacity or maximum intended load. The maximum intended load includes personnel, tools, accessory equipment, and work materials.

3.3.0 Personnel Platform Loading

The personnel platform shall not be loaded in excess of its rated load capacity. When a personnel platform does not have a rated load capacity, the personnel platform shall not be loaded in excess of its maximum intended load.

The number of employees, along with material, occupying the personnel platform shall not exceed the limit established for the platform and the rated load capacity or the maximum intended load.

Personnel platforms shall be used only for employees, their tools, and the materials necessary to do their work, and shall not be used to hoist only materials or tools when not hoisting personnel.

Materials and tools for use during a personnel lift shall be secured to prevent displacement. These items shall be evenly distributed within the confines of the platform while the platform is suspended.

3.4.0 Rigging

When a wire-rope **bridle sling** is used to connect the personnel platform to the load line, each bridle leg shall be connected to a master link or **shackle** in such a manner as to ensure that the load is evenly divided among the bridle legs (*Figures 2* and *3*).

Hooks on headache ball assemblies, lower load blocks, or other attachment assemblies shall be of a type that can be closed and locked, eliminating the hook throat opening. Alternatively, an alloy anchor type shackle with a bolt, nut, and retaining pin may be used.

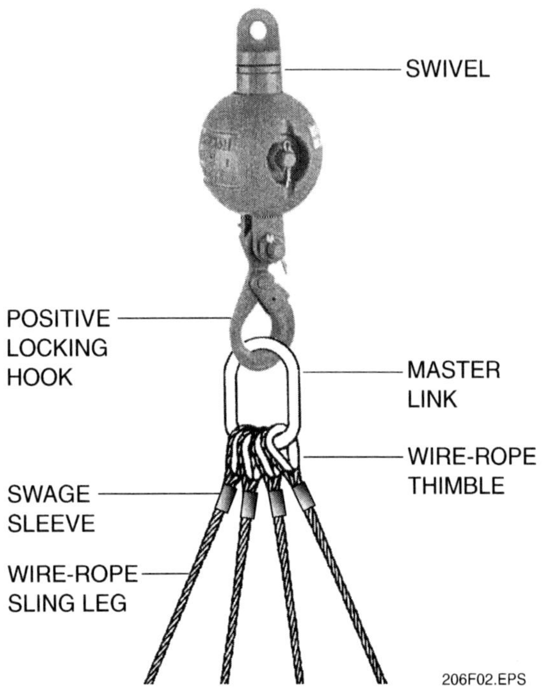

Figure 2 ◆ Bridle sling using a master link.

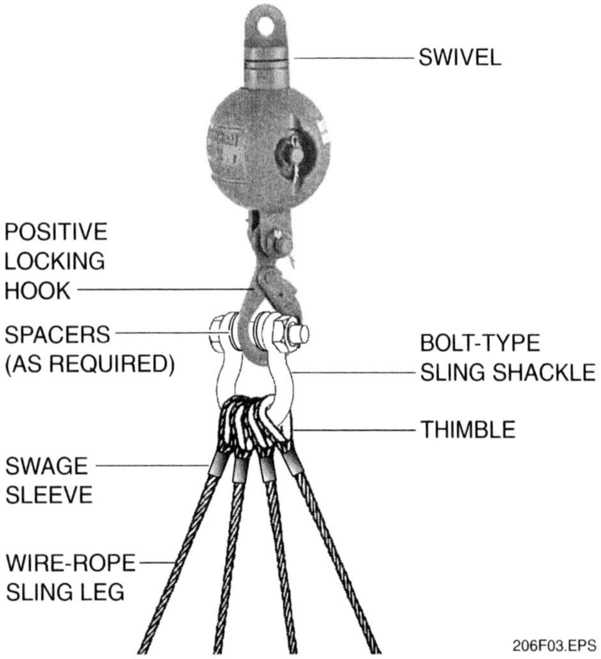

Figure 3 ◆ Bridle sling using a shackle.

Wire rope, shackles, rings, master links, and other rigging hardware must be capable of supporting, without failure, at least five times the maximum intended load applied or transmitted to that component. Where rotation-resistant rope is used, the slings shall be capable of supporting without failure at least ten times the maximum intended load. All eyes in wire rope slings shall be fabricated with thimbles.

Bridles and associated rigging for attaching the personnel platform to the hoist line shall be used only for the platform and the necessary employees, their tools, and the materials necessary to do their work, and they shall not be used for any other purpose when not hoisting personnel.

4.0.0 ◆ BOOM-ATTACHED PLATFORM REQUIREMENTS

Design criteria (*ASME B30.23*) for boom-attached platforms include the following:

- Based on the minimum yield strength of the material, the structural components of the platform responsible for supporting the load shall have a minimum design factor of two.
- Based on the minimum ultimate strength of the materials used, the structural components of the platform that are made of nonductile material and supporting the load shall have a minimum design factor of five.

Mandatory guidelines for specific features of personnel lifting platforms are as follows:

- A minimum platform rating of 300 lbs (136 kg), including tools and materials
- Guardrail protection consisting of a top rail, intermediate rail, toe board, and lower barrier with the following characteristics:
  – *Top rail:* Not less than 39 inches nor more than 45 inches in height with respect to the platform floor
  – *Toe board:* Around the entire perimeter of the platform and a minimum of $3\frac{1}{2}$ inches in height from the top of the platform floor
  – *Lower barrier:* Spanning the distance between the toe board and intermediate rail; it will also be of solid construction or expanded metal, having no opening that would allow a ½-inch sphere to pass through
- The top rail, toe board, and lower barrier must also be capable of withstanding a load of 300 pounds applied at any point in an outward or downward direction. This must occur without the rail, board, or barrier reaching its **yield** point.
- Anchorage points within the platform or on the boom tip which are suitable for attaching lanyards and capable of withstanding the impact created by falling loads.
- Hand railings, with a clearance of not less than 1½ inches between the railing and any other structure, located within the platform.
- Flooring with a slip-resistant surface and provisions to promote the free drainage of liquids.
- A method to secure loose items within the platform.

- An identification plate stating:
  - Manufacturer name and address
  - Platform rating in terms of weight and personnel
  - Platform identification number
  - Suspension system description
  - Weight of the empty platform and its suspension system
  - Date of the platform manufacture
  - Certification of compliance with the design, construction, and testing requirements of the *ASME B30.23* volume
  - Listing of any unique operational environments for which the platform has been designed
- Access gates, if used, equipped with a positive-acting device to suppress the gate from accidental opening.
- Minimal tipping due to movement of personnel on the platform.
- Minimum characteristics for the suspension system on a platform are as follows:
  - Rope sling systems with mechanically spliced flemish eyes designed with thimbles in the eyes
  - No wire-rope clips, wedge sockets or knots, synthetic webbing, or natural- or synthetic-fiber rope slings
  - Rope sling systems with poured socket end connections designed in accordance with the manufacturer's or qualified person's application instructions
  - Sling systems that utilize a master link or safety-type shackle for attachment to the hoisting equipment hook
  - For rope slings systems, each leg of the system permanently marked with the rated load of the leg
  - The master link in the system permanently marked with the suspension system's rated load and identification as a personnel lifting platform suspension component
  - Shackles that are categorized as a safety type
  - Overhead protection that allows view of hoist equipment components positioned above the personnel
  - Accessories for platform use in special environments included as specified by the manufacturer
  - Welds in accordance with the American Welding Society (AWS) or ASME
  - Bolted connections in accordance with the American Institute of Steel Construction, Inc. (AISC) *Specification for Structural Joints*
  - Weatherproof compartment for storage of the operator's manual and other documents

If motion controls are installed, they must be clearly marked, accessible to the operator, and protected from inadvertent actuation, and they must return to their neutral position and stop all motion when released. Boom motion controls, if installed, must be able to override platform controls. They also must have an emergency stop control and a control that is continuously activated for the controls to be operational.

Hoisting equipment used with personnel platforms shall meet these conditions:

- Have been removed for the period of personnel lifting if it has live boom or live load line capabilities
- Have an anti-two-block device, boom length indicator, positively locking hooks, and boom angle indicator
- Have automatic brakes such that when the equipment operating controls are released, the motions are brought to rest
- In the case of a holding device, be provided with hydraulic or pneumatic systems to prevent uncontrolled movement of the hoisting equipment in the event of a system failure
- Have a means to prevent the retraction of any hydraulically or pneumatically activated outriggers or stabilizers in the event a hydraulic or pneumatic line fails

5.0.0 ◆ CRANE AND OPERATIONAL REQUIREMENTS

According to *OSHA 29 CFR 1926.550(g)*, crane and operational requirements are broken down into operational criteria [*29 CFR 1926.550(g)(3)(i)*]; instruments and components [*29 CFR 1926.550(g)(3)(ii)*]; work practices [*29 CFR 1926.550(g)(6)*]; and traveling [*29 CFR 1926.550(g)(7)*]. These requirements are given in detail in the following sections.

5.1.0 Operational Criteria

Hoisting of the personnel platform shall be performed in a slow, controlled, cautious manner with no sudden movements of the crane or derrick or the platform.

Load lines shall be capable of supporting, without failure, at least seven times the maximum intended load. Where rotation-resistant rope is used, the lines shall be capable of supporting, without failure, at least ten times the maximum intended load. The required design factor is achieved by taking the current safety factor of 3.5, which is required under *OSHA 1926.550(b)(2)*, and applying the 50-percent derating of the crane capacity which is required by *OSHA 1926.550(g)(3)(i)(f)*.

Load and boom hoist drum brakes, swing brakes, and locking devices such as pawls or dogs shall be engaged when the occupied personnel platform is in a stationary position.

The crane shall be uniformly level within one percent of level grade and located on firm footing. Cranes equipped with outriggers shall have them all fully deployed following manufacturer's specifications, insofar as applicable, when hoisting personnel.

The total weight of the loaded personnel platform and related rigging shall not exceed 50 percent of the rated capacity for the radius and configuration of the crane or derrick.

The use of machines having live booms (booms in which lowering is controlled by a brake without aid from other devices which slow the lowering speeds) is prohibited.

5.2.0 Instruments and Components

Cranes and derricks with variable-angle booms shall be equipped with a boom angle indicator, readily visible to the operator.

Cranes with telescoping booms shall be equipped with a device to indicate clearly to the operator, at all times, the boom's extended length, or an accurate determination of the load radius to be used during the lift shall be made prior to hoisting personnel.

A positive-acting device shall be used which prevents contact between the load block or overhaul ball and the boom tip (anti-two-block device), or a system shall be used which deactivates the hoisting action before damage occurs in the event of a two-blocking situation (two-block damage prevention feature).

The load-line hoist drum shall have a system or device on the power train, other than the load hoist brake, which regulates the lowering rate of speed of the hoist mechanism (controlled load lowering). Free fall is prohibited.

5.3.0 Work Practices

The following work practices must be observed when using a suspended personnel platform. As defined in *29 CFR 1926.550(g)(3)(i)*, these requirements are detailed as follows:

- Personnel shall keep all parts of the body inside the platform during raising, lowering, and positioning. This provision does not apply to an occupant of the platform performing the duties of a signal person.
- Before workers exit or enter a hoisted personnel platform that is not landed, the platform shall be secured to the structure where the work is to be performed, unless securing to the structure creates an unsafe situation.
- Tag lines shall be used unless their use creates an unsafe situation.
- The crane or derrick operator shall remain at the controls at all times when the crane engine is running and the platform is occupied.
- Hoisting of personnel shall be promptly discontinued upon indication of any dangerous weather conditions or other impending danger.
- Personnel being hoisted shall remain in continuous sight of, and in direct communication with, the operator or signal person. In those situations where direct visual contact with the operator is not possible, and the use of a signal person would create a great hazard for the person, direct communication alone (such as by radio) may be used.
- Except over water, workers occupying the personnel platform shall use a body belt/harness system with lanyard appropriately attached to the lower load block or headache ball, or to a structural member within the personnel platform capable of supporting a fall impact for personnel using the anchorage. When working over water, the requirements of *OSHA 1926.106* shall apply.
- No lifts shall be made on another of the crane's or derrick's loadlines while personnel are suspended on a platform.

5.3.1 Working Over or Near Water – OSHA 1926.106

- Employees working over or near water where the danger of drowning exists shall be provided with U.S. Coast Guard-approved life jackets or buoyant work vests.
- Prior to and after each use, the buoyant work vests or life preservers shall be inspected for defects which would alter their strength or buoyancy. Defective units shall not be used.
- Ring buoys with at least 90 feet of line shall be provided and readily available for emergency rescue operations. Distance between ring buoys shall not exceed 200 feet.
- At least one lifesaving skiff shall be immediately available at locations where personnel are working over (or adjacent to) water.

5.4.0 Traveling

Hoisting of personnel while the crane is traveling is prohibited, except for portal, tower, and locomotive cranes, or where the supervisor demonstrates that there is no less hazardous way to perform the work.

Under any circumstances where a crane would travel while hoisting personnel, the employer or supervisor shall implement the following procedures to safeguard personnel:

- Crane travel shall be restricted to a fixed track or runway
- Travel shall be limited to the load radius of the boom used during the lift
- The boom must be parallel to the direction of travel
- A complete trial run shall be performed to test the route of travel before employees are allowed to occupy the platform.
- If travel is done with a rubber-tired carrier, the condition and air pressure of the tires shall be checked. The chart capacity for personnel lifts on rubber shall be used for application of the 50-percent reduction of rated capacity.
- Crane travel speed must be as slow as possible, and sudden stops and starts should be avoided.

6.0.0 ◆ PERSONNEL PLATFORM INSPECTION

According to *OSHA 29 CFR 1925.550(g)* guidelines, a visual inspection of the crane or derrick, rigging, personnel platform, and the crane or derrick base support or ground shall be conducted by a competent person immediately after the trial lift with the platform suspended several inches. This inspection is made to determine whether the testing has exposed any defect or produced any adverse effect upon any component or structure. Personnel shall not be hoisted unless the following conditions are determined to exist:

- Hoist ropes shall be free of kinks.
- Multiple-part lines shall not be twisted around each other.
- The primary attachment shall be centered over the platform.
- Fall protection equipment must be free of damage and defects.
- The hoisting system shall be inspected if the load rope is slack to ensure that all ropes are properly seated on drums and in sheaves.

Any defects found during inspections which create a safety hazard shall be corrected before hoisting personnel.

6.1.0 ASME-Prescribed Personnel Platform Inspection

ASME B30.23 guidelines break down inspections into three categories: initial, frequent, and periodic. Prior to use at each new job site, all platforms shall be inspected by a qualified person to verify compliance with the provisions of *ASME B30.23*. This is the initial inspection.

During a frequent inspection, the platform, suspension system, and attachment points shall be inspected by a designated person. This inspection is made to identify conditions that have been specifically indicated by the platform manufacturer or a qualified person designated by the employer as potentially creating a hazardous operating condition. The frequent inspection shall address at least the following items:

- Markings
- Structure
- Attachment mechanisms
- Special purpose items

Periodic inspections of the same items are to be conducted at least once every 12 months. Platforms that have been out of service for 12 or more consecutive months shall receive a periodic inspection prior to use.

Dated records of frequent inspections must be made and kept by the platform user for the duration of the personnel lift operation. Dated records of periodic inspections must be made by the platform owner. The last inspection shall be kept with the platform and be available for review.

7.0.0 ◆ TRIAL LIFT

The requirements for conducting a trial lift under *OSHA 29 CFR 1926.550(g)(5)* are as follows:

- A trial lift with the unoccupied personnel platform loaded at least to the anticipated liftweight shall be made from ground level or any other location where personnel will enter the platform, to each location at which the personnel platform is to be hoisted and positioned. This trial lift shall be performed immediately prior to placing personnel on the platform. The operator shall determine that all systems, controls, and safety devices are activated and functioning properly; that no interferences exist; and that all configurations necessary to reach those work locations will allow the operator to remain under the 50-percent limit of the hoist's rated capacity. A single trial lift may be performed at one time for all locations that are to be reached from a single setup position.

- The trial lift shall be repeated prior to hoisting personnel whenever the crane or derrick is moved and set up in a new location or returned to a previously used location. Additionally, the trial lift shall be repeated when the lift route is changed, unless the operator determines that the route change is not significant.
- At each job site, prior to hoisting personnel, the platform and rigging shall be proof-tested to 125 percent of the platform's rated capacity. This is accomplished by holding the platform in a suspended position for five minutes with the test load evenly distributed on the platform. Proof testing can be accomplished concurrently with the trial lift. After proof testing the platform, a competent person shall inspect the platform and rigging. If any deficiencies are found, they shall be corrected and another test conducted. Personnel hoisting shall not take place until the proof testing requirements are satisfied.

7.1.0 Proof Testing

In proof testing per *ASME B30.23*, the platform and rigging shall be proof-tested to 125 percent of the platform's rating. For this test, the following guidelines will be met:

- The load will be evenly distributed.
- The platform shall be hoisted and held in a suspended position for not less than five minutes.
- Inspection will follow after the proof testing.
- If damage is found, it shall be corrected and another proof test conducted.
- The hoisting equipment that will be used during the personnel lifts should be the equipment used to perform the proof tests.

After any structural repair or modification, the platform shall be tested at 150 percent of its rating. The platform shall be raised to a height, then lowered at a speed of 100 ft/min. Once 100 ft/min is reached, the platform descent will be halted by applying the hoisting equipment brakes. The platform shall be held for no less than 5 minutes. After the test, a qualified person will inspect the platform. If necessary, any damage revealed shall be corrected and the testing done again. The most recent record of proof testing shall be maintained at the job site.

8.0.0 ◆ ADVANCED OPERATIONS TECHNIQUES FOR HOISTING PERSONNEL

Lifting personnel near power lines is not allowed unless there is no less hazardous way to do the job. There are two options when hoisting personnel near power lines. The first option is best and involves de-energized and grounded power lines. In this situation, a representative from the power company or the owner of the power lines will ensure that the lines have been de-energized before hoisting takes place. Even with the power lines de-energized, signs must be installed at the operator's station, and on the outside of the crane, warning of electrocution or bodily injury unless minimum clearance (*Table 1*) is maintained.

WARNING!
Even if operator aids such as high-voltage proximity warning devices are being used, they do not negate the need for power-line shutdown.

The second, less desirable situation is when the power lines are energized, with the equipment outside the prohibited zone (*Figure 4*). In this situation, the possibility exists that the equipment or platform will become energized. A power line awareness checklist can be performed (see *Appendix C*). A pre-lift meeting must be held between the job-site supervisor and a representative of the electric company. Considerations must

Table 1 High-Voltage Power Line Clearances

CRANE IN OPERATION (1)	
POWER LINE (kV)	BOOM OR MAST MINIMUM CLEARANCES (feet)
0 to 50	10
50 to 200	15
200 to 350	20
350 to 500	25
500 to 750	35
750 to 1000	45
CRANE IN TRANSIT (with no load and the boom or mast lowered) (2)	
POWER LINE (kV)	BOOM OR MAST MINIMUM CLEARANCES (feet)
0 to 0.75	4
0.75 to 50	6
50 to 345	10
345 to 7500	16
750 to 1000	20

Note 1: For voltages over 50kV, clearance increases 5 feet for every 150 kV.
Note 2: Environmental conditions such as fog, smoke, or precipitation may require increased clearances.

206T01.EPS

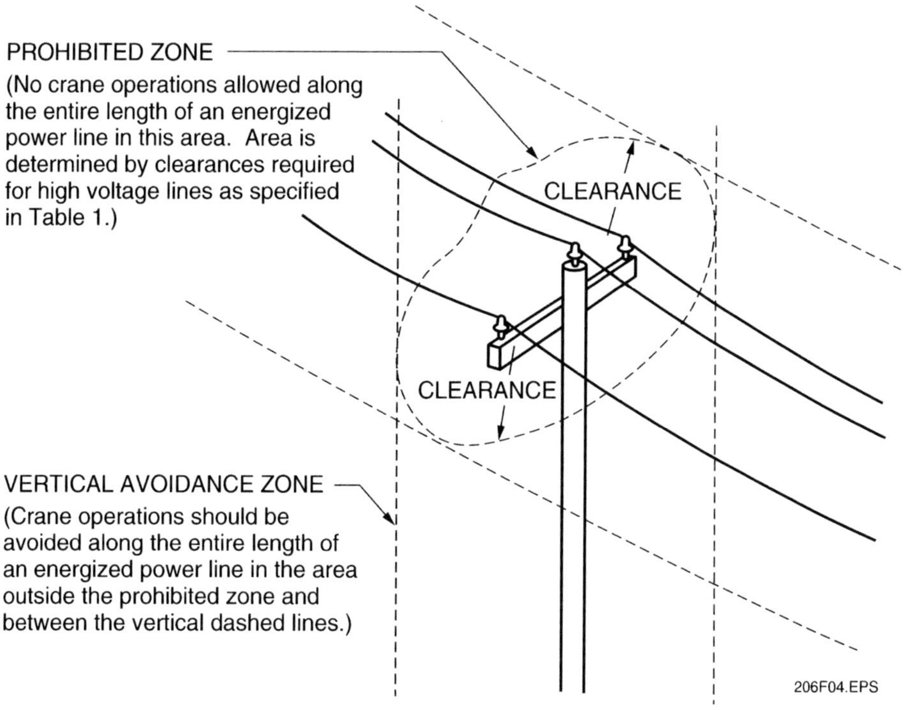

Figure 4 ◆ Prohibited zone and avoidance zone.

be given to possible power-line movement and platform restraint devices. Signs must be made available to warn of possible electrocution and bodily injury. These signs are to be posted at the operator station, on the outside of the equipment, and inside the personnel platform. In some cases, large high-voltage warning signs with upward pointing arrows placed under the lowest portion of the energized line may be desired or required.

The final situation is when the equipment must be moved under the prohibited zone and through the avoidance zone. In this situation, the boom must be lowered (*Figure 5*) and no personnel are permitted in the platform. During transit, the minimum clearances (*Table 1*) from power lines must be obeyed, and any speed and terrain effects on boom and equipment movement must be taken into consideration.

Figure 5 ◆ Moving a crane through an avoidance zone.

Summary

Although using cranes to lift personnel was common in the past, OSHA regulations prohibit the practice unless no other method is suitable. When lifting is permitted, certain controls must be in place. These controls are covered under *OSHA 29 CFR 1926.104, 1926.106, 1926.550(g), 1926.753*, and *ASME B30.23.*

The purpose of this module is to cover in-depth the guidelines of both the federal regulations and the consensus standard. The module discusses fall protection, platform requirements, crane operational requirements, how to inspect personnel platforms, and how to hoist personnel near power lines.

Review Questions

1. A small cord or rope used for securing or suspending is a _____.
   a. tag line
   b. swage sleeve
   c. rope bridle
   d. lanyard

2. The proper place for securing a lanyard is a(n) _____.
   a. shackle
   b. anchorage
   c. guardrail
   d. swage sleeve

3. A full-body harness lanyard must be a minimum of _____ rope or equivalent.
   a. ½-inch nylon
   b. ¾-inch wire-core manila
   c. ¾-inch nylon
   d. ⅞-inch wire-core manila

4. Lifelines used in areas where they may be subjected to cutting or abrasion must be a minimum of _____ rope.
   a. ½-inch nylon
   b. ¾-inch wire-core manila
   c. ¾-inch nylon
   d. ⅞-inch wire-core manila

5. A personnel platform and suspension system is also known as a manbasket.
   a. True
   b. False

6. A personnel platform must be capable of supporting its own weight and at least _____ times the maximum intended load.
   a. two
   b. three
   c. four
   d. five

7. A personnel platform must have a guardrail, be enclosed from toeboard to mid-rail, and have a _____.
   a. grab rail
   b. locking device
   c. elastic floor
   d. lightning device

8. The platform must have a permanent marking stating the _____.
   a. type of crane that can lift it
   b. weight of the platform and the rated load capacity
   c. year it was made
   d. wind speeds that the platform can withstand before operations should cease

9. A grab rail must be installed _____.
   a. on each corner of the platform
   b. inside the entire perimeter of the platform
   c. on the floorboard
   d. at the owner's discretion

10. According to *OSHA 1926.550(g)*, access gates must be equipped with a _____.
    a. restraining device
    b. 5-inch handle
    c. 10-inch handle, for easy gripping
    d. steel grip device

11. The number of employees occupying the platform must not exceed _____.
    a. the limit established for the platform
    b. two (the signal person and the operator)
    c. three
    d. four

12. According to *OSHA 1926.550(g)*, each bridle-sling leg must be connected to a _____ or shackle to ensure the load is evenly distributed.
    a. rod
    b. lanyard
    c. master link
    d. auxiliary link

13. According to *ASME B30.23*, the toeboard on the guardrail must be a minimum of _____ in height from the top of the platform floor.
    a. 3½ inches
    b. 6 inches
    c. 1 foot
    d 2 feet

14. Welds on the suspension system must be in accordance with ASME or _____.
    a. CMAA
    b. DOE
    c. NEC
    d. AWS

15. The top rail on a personnel platform must be not less than _____ in height with respect to the platform floor.
   a. 24 inches nor more than 48 inches
   b. 36 inches nor more than 60 inches
   c. 39 inches nor more than 45 inches
   d. 48 inches nor more than 66 inches

16. The lower barrier of the guardrail on a personnel platform spans the distance between the _____.
   a. top rail and the toe board
   b. grab rail and the intermediate rail
   c. top rail and the intermediate rail
   d. toe board and the intermediate rail

17. A personnel platform must have hand railings with a clearance of not less than _____ between the railing and any other structure.
   a. ½ inch
   b. 1 inch
   c. 1½ inches
   d. 3 inches

18. Cranes having _____ must *not* be used for hoisting personnel.
   a. crawlers for stabilizers
   b. live booms
   c. lattice booms
   d. hydraulic booms

19. Load lines must be capable of supporting, without failure, at least _____ times the maximum intended load.
   a. 4
   b. 5
   c. 6
   d. 7

20. Controlled load lowering is described as lowering that _____.
   a. uses positively locking hooks
   b. uses tag lines
   c. is regulated by a system or device on the power train, other than the load hoist brake
   b. is controlled by a brake without aid from other devices that slow the lowering speed

21. Cranes and derricks with variable-angle booms must be equipped with a _____.
   a. control lockout system
   b. boom angle indicator
   c. proximity warning device
   d. hoist drum indicator

22. According to *OSHA 1926.550(g)*, when working over or near water the requirements of _____ must apply.
   a. *1910.104*
   b. *1910.114*
   c. *1926.104*
   d. *1926.106*

23. Dated records of _____ inspections must be made and kept by the platform user for the duration of the personnel lift operation.
   a. initial
   b. frequent
   c. periodic
   d. annual

24. Platforms which have been out of service for 12 or more consecutive months must receive a(n) _____ inspection prior to use.
   a. initial
   b. frequent
   c. periodic

25. After any structural repair or modification during proof testing, the platform must be tested at _____ percent of its rating.
   a. 120
   b. 130
   c. 140
   d. 150

GLOSSARY

Trade Terms Introduced in This Module

Anchorage: A place for properly securing a lanyard.

Bridle sling: A multi-leg, wire-rope sling.

Guardrail: A system of protective railings at waist height, mid-height, and toeboard that provides an added measure of safety around platforms and catwalks.

Lanyard: Any of various small cords or ropes used for securing or suspending.

Shackle: A fastening or coupling device.

Yield: To give way to force, pressure, etc., so as to move, bend, or collapse.

APPENDIX A

Personnel Platform Lift Planning Authorization Form

Note: This Appendix is part of *ASME B30.23-1998*

1. Location: ______________________ Date: ______________________
2. Purpose of Lift: ______________________
3. Hoist Eqpt. Mfg:______________ Model #:______________ Serial #: ______________
4. Expected Radius:______________ (maximum);______________(at work location)
5. (A) Rated Load at Radius:______________ (B) Maximum Lift Load:______________[50% of 5(A)]
6. Platform ID:______________________ Platform Rating: ______________
7. Platform Weight:______________________ Type: (Pin On)________(Suspended)
8. (A) Number of Platform Occupants:______________ (B) Approx. Wt. (With Equip): ________
9. Total Lift Weight:______________ ______________[7 + 8(B)] [No more then 5(b) above]
10. Personnel Lift Supervisor: ______________________
11. What Are The Alternatives To This Personnel Lift?______________________
12. Why Are They Not Being Used? ______________________
13. Pre-Lift Briefing Held: ______________ /________/________/ ______________AM/PM
    Attendees:______________________
14. Anticipated Hazards (wind, weather, visibility, power lines):______________________
15. Lift Accomplished Date: ______________________ Time:______________
16. Remarks: ______________________

____

Personnel Lift Authorizer Signature Date

APPENDIX B

Personnel Platform Pre-Lift Inspection

Note: This Appendix is part of *ASME B30.23-1998*

Inspector: ______________________________ Date: ____________

Platform ID: ______________________________

	SAT	UNSAT
1. Markings		
Platform (All Information Legible)	______	______
Suspension System	______	______
2. Structure		
Load Supporting Welds/Bolts	______	______
Load Supporting Members	______	______
Barrier From Toe Board to Intermediate Rail	______	______
Hand Rail	______	______
Fall Protection Device Anchorage Points	______	______
Gate Locking Mechanisms	______	______
Platform Flooring	______	______
Suspension Attachment Points	______	______
3. Attachment Mechanisms		
Pins/Ears/Bolt-Ups/Eyes (Circle)	______	______
Wire Rope/Chain/Rigid Leg (Circle)	______	______
Master Links	______	______
4. Special Purpose Items		
(Overhead Protection, Flotation, Platform Controls)		
List: (1) ______________________	______	______
(2) ______________________	______	______
(3) ______________________	______	______

5. General Comments: ______________________________

____

Personnel Lift Authorizer Signature Date

APPENDIX C

Power Line Awareness Permit

POWER LINE AWARENESS PERMIT

Today's Date __________ **Job Number** ______________________________

Contractor Name			
Job Address			
Telephone Number		**Fax Number**	

Emergency Contact Number	

Survey

Before beginning any project, you must first survey your work area to find power lines at the job site. (See job-site sketch on reverse side.)

Identify

After finding all of the power lines at your site, identify the activities you ll be doing that may put you or your workers at risk. Mark one or more of the following:

- ☐ Cranes (mobile or truck mounted)
- ☐ Drilling rigs
- ☐ Backhoes/excavators
- ☐ Long-handled tools
- ☐ Other tools/high-reaching equipment
- ☐ Concrete pumper
- ☐ Aerial lifts
- ☐ Dump trucks
- ☐ Ladders
- ☐ Material handling & storage
- ☐ Scaffolding
- ☐ Other__________________

Eliminate or Control

After identifying the power line and high-risk activities on the job site, determine how to eliminate or control the risk of electrocution (a successful determination is often reached only after consultation with the utility).

Mark one or more of the following:

- ☐ Move the activity
- ☐ Change the activity
- ☐ Have the utility de-energize the power line
- ☐ Have the utility move the power line
- ☐ Use barrier protection (insulated sleeves)
- ☐ Use an observer
- ☐ Use warning lines with flags
- ☐ Use non-conductive tools
- ☐ Use a protective technology:
  - ☐ Insulated link
  - ☐ Boom cage guard
  - ☐ Proximity device

Always maintain your minimum safe clearance distance from the power line, except when the utility has de-energized and visibly grounded the power line.

Voltages	**Distance from Power Line**
Less than 50 kV	10 Feet
More than 50 kV	10'+(0.4")(# of kV over 50 kV)

WARNING!
It is unlawful to operate any piece of equipment within 10' of energized lines

CONSTRUCTION SAFETY COUNCIL

206A01.EPS

Job-Site Sketch

(Draw in location of power lines and their proximity to construction site, including such things as proposed excavations, location of heavy equipment, scaffolding, material storage areas, etc.)

Completed by____**Date**______________

Title____

Approved by____**Date**______________

Title____

CONSTRUCTION SAFETY COUNCIL

206A02.EPS

Figure Credits

Lifting Technologies, Inc.	206F01
The Crosby Group, Inc.	206F02 (photo), 206F03 (photo)
Wire Rope Technical Board	206F02 (art), 206F03 (art)
Construction Safety Council	206F05, Appendix C
Reprinted from ASME *B30.23-1998* by permission of The American Society of Mechanical Engineers. All rights reserved.	Appendix A and B

CONTREN® LEARNING SERIES — USER UPDATE

The NCCER makes every effort to keep these textbooks up-to-date and free of technical errors. We appreciate your help in this process. If you have an idea for improving this textbook, or if you find an error, a typographical mistake, or an inaccuracy in NCCER's Contren® textbooks, please write us, using this form or a photocopy. Be sure to include the exact module number, page number, a detailed description, and the correction, if applicable. Your input will be brought to the attention of the Technical Review Committee. Thank you for your assistance.

Instructors – If you found that additional materials were necessary in order to teach this module effectively, please let us know so that we may include them in the Equipment/Materials list in the Annotated Instructor's Guide.

Write: Product Development and Revision
National Center for Construction Education and Research
P.O. Box 141104, Gainesville, FL 32614-1104

Fax: 352-334-0932

E-mail: curriculum@nccer.org

Craft ____________________ Module Name ____________________

Copyright Date __________ Module Number __________ Page Number(s) __________

Description

____

____

____

____

(Optional) Correction

____

____

____

(Optional) Your Name and Address

____

____

____

Index

Index

Abrasion
of sling, 5.17
of wire rope, 1.8, 1.9, 1.10, 1.19
Acceleration, 5.27
Accidents
equipment failures, 6.5
falling objects, 3.5, 3.6, 3.12, 3.13, 5.13, 6.3
falls, 6.2, 6.6
from electrical hazards, 6.8–6.9
jacking operations, 5.13
pinching, 5.15
structural failure of crane, 5.9, 5.24
tipping of crane, 2.1, 3.5, 5.7–5.9, 5.24
tipping of personnel platform, 6.2
Administration, 4.15
American Institute of Steel Construction, Inc. (AISC), 6.5
American National Standards Institute (ANSI), 5.21
American Society of Mechanical Engineers (ASME)
ANSI/ASME spreader beam standards, 5.21
hand, voice, and audible signal standards, 5.24
personnel platform standards, 6.2, 6.4, 6.5, 6.7, 6.8
prohibited zone for power lines, 6.8–6.9
American Welding Society (AWS), 6.5
Anchorage point, 6.2, 6.4, 6.6, 6.13
Anemometer, 2.6
Annealed, 1.9, 1.19
ANSI. *See* American National Standards Institute
Anti-two-blocking device, 6.1, 6.5, 6.6
ASME. *See* American Society of Mechanical Engineers
Avoidance zone, 6.9

Backstop, 3.8
Bail, 5.32
Bar
lever, 5.15
spreader, 3.10, 3.17, 5.22, 5.33, 5.34. *See also* Beam, spreader
Barricades, 3.4
Base
boom, 3.6, 3.7, 3.9
crane, 5.7
jib, 3.12
Beam(s)
adjustable, 5.22–5.23
equalizer, 4.5, 4.6, 4.10, 4.11, 4.12, 4.13, 5.23
overview on use, 5.21–5.22
safe working load during cribbing, 5.15
spreader, 5.21–5.23. *See also* Bar, spreader
Bearing points, 2.12
Becket, 2.7, 2.9, 2.10, 2.21
Becket assembly, 1.5–1.6
Bill of lading, 5.23
Birdcaging of wire rope, 1.7, 1.8, 1.9, 1.19
Block(s). *See also* Load block
assembly / disassembly, 3.3, 3.8, 3.11, 3.12, 3.13
for the crawler idler, 3.9, 3.12
hook, 4.5, 4.6, 4.7
with jack, 5.13, 5.15
overview, 5.15
traveling, 2.16
Blocking, 2.1, 5.23
Bob's Rigging and Crane Handbook, 5.19
Bolts, 1.7, 6.3
Boom(s)
Chicago, 5.23
components, 3.5
lattice, 3.5–3.13
live, 6.6
motion controls, 6.5
power line clearance for, 6.8
raising or lowering, 3.11, 3.13, 5.7
removal or stowage, 2.6, 2.10–2.11, 2.12
retraction, 2.7, 2.10, 2.12, 2.13, 2.15, 5.7
sections
assembly, 3.6–3.8, 3.9–3.11
installation of intermediate, 2.11–2.12
removal of intermediate, 2.12–2.13
sudden deflection, 5.24
sudden stop, 5.9
telescoping attachment. *See* Boom extension
Boom angle
and crane stability, 5.9
indicator, 3.8, 6.1, 6.5
in load / capacity charts, 4.3, 4.4
in multiple-crane lifts, 4.6
Boom extension
assembly / disassembly, 2.1–2.6
during critical lifts, 6.6
intermediate boom sections, 2.11–2.13
during personnel lifts, 6.6
power-pinned fly, 2.13–2.15
Boom head, auxiliary single-sheave, 2.15–2.16

Boom length
indicator, 6.1
in load / capacity charts, 4.3, 4.4
in multiple-crane lifts, 4.6
Boom nose, 2.3, 2.5, 2.7, 2.8, 2.10, 2.11
Boom point elevation, 4.3, 4.4
Boom stop, 3.8, 4.5
Boom tip, end termination of wire rope, 1.4
Boom toe, 2.13, 2.15
Box
channel, 5.31
electrical distribution, 2.13
Bracket(s)
front and rear support, 2.3, 2.4, 2.5, 2.7, 2.11
intermediate boom section, 2.13
jib, 3.12
stowage, 2.10
Brake
crane, 6.1
hoist, 6.1, 6.6
personnel platform, 6.5, 6.6, 6.8
Bundles, rebar, 5.21, 5.23–5.24
Buoy, 6.6
Buoyancy, 5.24, 5.27
Burring, of wire rope, 1.10
Bushing, retainer, 2.7, 2.8, 2.10, 2.21

Cable, 2.8, 2.9, 2.10. *See also* Hoist rope or cable or line; Wire rope
Calculations
engineering considerations, 4.14–4.15
lift plan with a single crane, 4.2–4.5
lift plan with multiple cranes, 4.6–4.13
load balance, 5.4–5.5
load for vertical vessel up-ending and setting, 5.34–5.35
pull needed on an incline plane, 5.16
safe working load, 5.19, 5.20
Cantilever, 3.9, 3.10, 3.17
Capacity, 6.1, 6.3, 6.6. *See also* Load / capacity chart; Sling capacity chart
Carbody, 3.3
Carrier body, 5.7
Catwalk, 4.5
Center of gravity
on a crane, 5.5, 5.7–5.9
definition, 5.27
of the load, 5.5, 5.7, 5.9, 5.15, 5.24
during multiple-crane lifts, 4.5, 4.6, 4.7–4.10
and rotational forces, 5.3–5.4, 5.5
Center of rotation, 5.6, 5.8
Chocking, 5.15, 5.27
Choke angle, 5.20, 5.21, 5.24
Choker, 4.12, 4.13, 5.18, 5.24, 5.33
Choker hitch arrangement. *See* Hitch, choker
Chord, 3.5, 3.17
Clamp, beam, 4.7, 4.8, 5.31
Clearance, 6.8
Clips
retainer
in A-frame jib installation, 2.8, 2.9
in A-frame jib stowage, 2.10
in auxiliary single-sheave boom head installation, 2.16
in boom section installation, 2.12
in boom section removal, 2.13
in telescopic boom installation, 2.3, 2.5, 2.6
wire rope, 1.5, 1.7, 5.17
Code of Federal Regulations, 5.16
Comealong, 5.15
Communication
during crane assembly / disassembly, 3.4
hand signals, 4.6, 4.14, 4.15, 5.23
for implementation of lift plan, 4.14
during multiple-crane lift, 4.6
overview, 5.24–5.25
during personnel lift, 6.6
pre-lift meeting, 4.15, 6.8
radio, 4.6, 4.15
Configuration
hitch arrangement, 5.19
in the lift plan, 4.31, 4.32, 4.33
Core, of wire rope
components, 1.1, 1.2, 1.3, 1.4
definition, 1.2, 1.19
failure, 1.7, 1.9
Corrosion, of wire rope, 1.7, 1.8, 1.19
Counterweight(s)
clearance, 2.1
and crane center of gravity, 5.7, 5.8
front bumper, 3.9
during lattice boom crane assembly / disassembly, 3.2, 3.3, 3.4–3.5, 3.6, 3.9
requirements, 2.1, 3.4
use of lifting capacity table, 2.11
Crane
crawler, 5.6, 5.7
hydraulic, 5.23
tailing, 5.33
truck, 5.23
Crane operator, 3.4, 4.6, 4.14, 5.9, 5.25
Crank, hand, 2.14, 2.15
Cribbing, 5.15, 5.23
Crimping, of shank to wire rope, 1.5
Crowning, of wire rope, 1.10
Crushing, of wire rope, 1.7, 1.10
Curling, of wire rope, 1.10
Cylinder, jack, 5.11, 5.12, 5.13

Data
identification, 1.4, 3.6, 3.7, 3.8
safe working load and tare weight, 5.21
D/d ratio, 1.10, 1.11, 5.17, 5.19, 5.33
Dead end, 1.6, 1.12, 1.19
Derigging plan, 4.2
Design factor, 1.12–1.13, 1.19
Documentation
bill of lading and the shipping list, 5.23
critical lift plan form, 4.27–4.33
inspection records, 1.7
maintenance records, 1.10
personnel platform inspection form, 6.7, 6.16
personnel platform lift plan form, 6.15
power line awareness checklist, 6.8, 6.17–6.18
pre-raising checklist, 3.11
Dog, 6.1, 6.6
Dolly, 5.15
Dragline fairlead, 4.5
Drawings
boom rigging, 3.5
in the lift plan, 4.2, 4.6, 4.7
rigging sketch, 4.14
Drum(s)
correct rope cycle on, 2.16, 3.8
crossover of wire rope on, 1.9
definition, 1.19
function, 1.13
OSHA regulations, 6.1, 6.6

reeving technique, 1.13–1.14, 2.16
transmission of power to, 1.1
types, 1.13

EEIPS. *See* Extra extra improved plow steel
Efficiency, of wire rope end termination, 1.4, 1.19
EIPS. *See* Extra improved plow steel
Electrical hazards. *See* Power lines
Electric arc, 1.9
Elevation. *See* Boom point elevation
Emergency procedures
power lines, 4.15, 6.8–6.9, 6.17–6.18
signals, 5.25
Engineering considerations, 4.14–4.15, 4.19, 5.1, 5.21
Equalizer, 4.19
Equipment
hoisting, 5.23–5.24
jacks, 3.3, 5.11–5.13
Extractor, for tube bundle, 5.31–5.32
Extra extra improved plow steel (EEIPS), 1.2
Extra improved plow steel (EIPS), 1.2
Eyebolt, 5.21
Eye of a sling, 5.16, 5.17, 5.18, 5.33

Failures, 5.9, 5.24, 6.5
Fall protection, 6.2, 6.6
Fatigue, of wire rope, 1.8, 1.9, 1.19
FC. *See* Fiber core
Fiber core (FC), 1.2, 1.5, 1.19
Finish, of wire rope, 1.4
Fittings
anchor and attach, for swing-away extension, 2.3, 2.4, 2.5, 2.6
for harness and lanyard, 6.2
in jack system, 5.12
for platform rope sling, 6.5
wire rope end, 1.4
Fleet angle, 1.14–1.15
Flow divider, 2.13, 2.15
Fly (swing-away lattice extension), 2.2–2.6
Fly, power-pinned, 2.13–2.15
Fracture, fatigue, 1.9
Friction, 1.10, 1.12, 5.9, 5.16
Fulcrum, 5.3–5.5, 5.7

Gantry
definition, 3.17
extension, 3.6
in lattice boom assembly, 3.3, 3.5, 3.6
weight deduction for, 4.5
Gate, access to platform, 6.2, 6.5
Gauges, in jack system, 5.12
Gear reducer, 2.13, 2.14, 2.15
Grip hoist, 5.13
Ground
level, 6.1, 6.6
rough-terrain quadrants of operation, 5.6
stability, 4.14, 5.24
Grounding, electrical, 6.8
Guardrail, 6.2, 6.3, 6.4, 6.13
Guide, for cable or rope
for auxiliary single-sheave boom head, 2.16
for jib, 2.9, 2.10, 3.12
for telescopic boom, 2.3, 2.5, 2.6
weight deduction for, 4.5

Hand signals, 4.6, 4.14, 4.15, 5.23, 5.24–5.25
Hardware. *See* Bolts; Fittings; Nuts; Pins
Harness, full-body, 6.2, 6.6
Hazardous materials, 4.14
Headache ball, 1.5, 1.19, 2.6
Heat damage, 1.9
Heat exchanger, 5.1, 5.2, 5.21, 5.28–5.32
Hitch
back, 3.5
basket, 5.17, 5.19–5.20, 5.33
bridle, 5.19
choker, 5.20–5.21
Hoist(s)
drum. *See* Drum
grip, 5.13
power equipment for, 5.23–5.24
roller, 4.5
Hoist rope or cable or line
for A-frame jib assembly / stowage, 2.7, 2.8, 2.9, 2.10, 2.11
auxiliary, 2.3, 2.6, 2.7, 2.21
for auxiliary boom head installation, 2.16
for boom section installation / removal, 2.12, 2.13
for extension assembly, 2.3, 2.6
for lattice boom assembly / disassembly, 3.8, 3.11, 3.13
for multiple-crane lift, 4.6
for personnel lift, 6.1, 6.5, 6.7
weight deduction for, 4.5
Hook(s)
in boom section installation, 2.12
not recommended for personnel platforms, 6.3
positive-locking, 6.4
safety, on spreader beam, 5.23
wedge socket to attach, 1.5
Horns, 5.25
Hoses, in jack system, 5.12
Hydraulic equipment, 5.11, 5.12, 5.23, 6.5

Impact loading, 5.9
Improved plow steel (IPS), 1.2, 5.18
Independent wire rope core (IWRC), 1.2, 1.5, 1.19
Inspections
boom parts, 3.5, 3.13
personnel platform system, 6.7, 6.8, 6.16
records, 1.7
sheave and load block, 1.9–1.10
sling, 5.16
wedge socket components, 1.6
wire rope, 1.7–1.9
Installation
A-frame jib, 2.7–2.9, 2.11
auxiliary single-sheave boom head, 2.15–2.16
intermediate boom sections, 2.12
swing-away lattice extension, 2.2–2.6
tube bundle in a heat exchanger, 5.28–5.30
wire rope, 1.13–1.14, 2.16
Inventory, 3.5
IPS. *See* Improved plow steel
IWRC. *See* Independent wire rope core

Jacking system, 5.12
Jack operations, 5.13, 5.15
Jacks, 3.3, 5.11–5.13
Jib(s)
A-frame, 2.2, 2.3, 2.4, 2.6–2.11, 2.12
assembly, 3.12
attachments, 4.5
in the lift plan, 4.1
weight deduction for, 4.5
Jib adapter, 3.12
Jib angle, 5.7
Job planning, 1.10, 2.1

Kickout, 3.9
Kinks, in wire rope, 1.7, 1.8, 1.10, 1.19

Lacing, 3.5, 3.17
Lagging, 4.5
Lang lay, 1.2
Lanyard, 6.2, 6.6, 6.13
Lateral movement of load, 5.1, 5.2, 5.31–5.33
Lattice boom cranes. *See also* Boom extension
  boom assembly, 3.5–3.12
  boom disassembly, 3.12–3.13
  boom parts, 3.5
  counterweight considerations, 3.4–3.5
  crane assembly / disassembly procedures, 3.2–3.3
  jib assembly, 3.12
  overview, 3.1–3.5
  pre- and post-assembly considerations, 3.4
Lay (length), 1.2, 1.4, 1.19
Lay (strand), 1.2, 1.4, 1.13, 1.19
Legs. *See* Sling
Length, of wire rope, 1.2, 1.3
Level
  ground, 3.4, 3.12, 4.5, 4.6, 6.1, 6.6
  personnel platform, 6.2
Lever, 5.27
Leverage, 5.3, 5.5, 5.7–5.8, 5.27
Lifeline, 6.2
Lift operations
  critical, 4.14, 4.21–4.33
  with horizontal component, 5.1, 5.2, 5.9, 5.31–5.33
  multiple-crane, 4.5–4.13, 5.33–5.35
  in or near water, 5.21, 5.24, 6.6
  planning. *See* Lift plan
  practice or trial, 4.15, 6.1, 6.7–6.8
  up-ending and setting vertical vessels, 5.33–5.35
Lift plan
  calculations
    engineering considerations, 4.14–4.15
    with multiple cranes, 4.6–4.13, 5.33–5.35
    with a single crane, 4.2–4.5
  definition, 4.19
  example, 4.21–4.33
  implementation, 4.15
  overview, 4.1–4.2, 5.25
  for personnel lift, 6.15
  sign off, 4.15
Line(s)
  hydraulic or pneumatic, 6.5
  load, 4.3, 6.1, 6.5
  parts of a, 1.12–1.13
  reeving for. *See* Reeving
  restrainer, on downward slope, 5.15
  tag, 1.13, 4.15, 5.23, 6.6
  whip, 3.11, 4.3
  wire. *See* Wire rope
Link(s)
  attaching, 2.4, 2.7, 2.8, 2.10
  master, 6.3, 6.4, 6.5
  offset, 2.8, 2.9, 2.10
Live section, 1.6, 1.12, 1.19
LMI. *See* Load moment indicator
Load
  asymmetrical, 5.9, 5.19, 5.20
  in balance, 5.3, 5.4
  buoyancy, 5.24
  capacity calculations for total, 1.12–1.13
  center of gravity, 5.5, 5.7, 5.9, 5.15, 5.24
  damage by slings or chokers, 5.21
  dynamics, 5.1, 5.3–5.9
  engineering considerations, 4.14–4.15, 5.1
  lateral movement, 5.1, 5.2, 5.31–5.33
  maximum for personnel platform, 6.1, 6.3, 6.4, 6.6
  in multiple-crane lifts, 4.6, 4.10–4.13, 5.33–5.35
  oddly-shaped, 5.20–5.21
  percent capacity guidelines, 4.22
  on personnel platform, 6.1, 6.3
  reduction with reeving, 1.11
  sail area and force of wind, 5.9
  secured, 5.15
  sudden release, 5.9, 5.24
  tensile, for lanyard hardware, 6.2
  testing, 1.7
  unloading procedure, 5.17, 5.23
Load block
  in boom section installation, 2.12
  definition, 1.10, 1.19
  disassembly, 2.10
  end termination of wire rope, 1.4
  inspection, 1.9–1.10
  load and capacity markings, 1.9
  in multiple-crane lifts, 4.6
  twisting, 1.13
  wedge socket to attach, 1.5
Load / capacity chart
  Manitowoc 4100W-2, 4.3–4.5
  and parts of line, 1.12
  reduction of ratings with wind, 5.9
  use prior to boom installation, 2.11, 3.6, 3.8, 3.9
  use prior to moving load, 5.23
Load moment, 5.5, 5.7, 5.8
Load moment indicator (LMI), 2.2, 2.6, 2.12
Load operating radius
  in capacity charts, 4.3, 4.4
  and crane stability, 5.5, 5.9
  definition, 5.27
  for multiple-crane lifts, 4.6, 4.10
Lock, swing, 3.4, 3.6, 3.9, 3.12
Loop, 1.4, 1.5, 1.19
Lubrication, 1.10
Lug, 2.9, 2.10, 2.21, 3.5, 5.21, 5.33

Maintenance, 1.10, 2.16, 3.5. *See also* Inspections
Manitowoc 999 crane, 3.3
Manitowoc 4100W-2 crane, 4.3–4.5
Manufacturer's data and recommendations
  A-frame jib, 2.6, 2.7
  booms, 2.11, 3.5, 3.6, 3.8, 3.9, 3.12
  catalog nominal breaking strength, 1.12
  crane operator's manual, 3.5
  D/d ratio, 1.10
  lubricant, 1.10
  platform operator's manual, 6.5
  power-pinned fly, 2.13
  quadrants of operation, 5.6, 5.7
  swing-away lattice extension, 2.2
  wire rope specifications, 1.11
Mast
  A-frame jib, 2.8, 2.9, 2.10, 2.21
  boom, 3.6
  clearances for power lines, 6.8
  lattice boom crane, 3.3
Mast arm, 3.3
Mobile crane(s)
  auxiliary, 2.12, 2.13
  footprint, 5.6
  heavy-lift, 5.3

hydraulic telescoping, fly used with, 2.13–2.14
in the lift plan, 4.1, 4.2, 4.14, 4.31
Manitowoc 999, 3.3
Manitowoc 4100W-2, 4.3–4.5
placement on the work site, 3.4, 4.1
pre- and post-assembly considerations, 2.1, 3.4
quadrants of operation, 5.6
stability, 5.5–5.9, 5.10. *See also* Accidents, tipping; Tipping axis
structural failure, 5.9, 5.24
tail, 4.22
traveling, 6.7
use of multiple, 4.5–4.13, 5.33–5.35
Moments, 5.3–5.4
Monel™, 1.2, 1.19

Nuts, 1.7, 6.3

Occupational Safety and Health Administration (OSHA)
minimum wraps on drum, 1.13
mobile crane standards, 4.2
personnel platforms and lift operations, 6.1, 6.2, 6.5, 6.7, 6.9
Ocean operations, 5.21, 5.24, 6.6
OSHA. *See* Occupational Safety and Health Administration
Outrigger(s)
during boom extension assembly, 2.3, 2.5
during boom head assembly, 2.15
during boom section assembly, 2.11
during crane assembly / disassembly, 3.4
and crane center of gravity, 5.7
hydraulic or pneumatic line failure, 6.5
during jib assembly / disassembly, 2.7, 3.12
during personnel lift, 6.6
during power-pinned fly extension / retraction, 2.13, 2.15
quadrants of operation and crane stability, 5.6, 5.7

Pawl, 3.8, 3.10, 3.11, 6.1, 6.6
Peening, 1.9
Pendant
during boom assembly, 3.8, 3.10–3.11
during boom disassembly, 3.13
boom handling, 3.3, 5.9
jib forestay and backstay, 3.12
overview, 3.5, 3.17
Pendulum effect, 5.9, 5.27
Personal protection equipment (PPE)
assembly and disassembly of crane, 2.2, 3.4
fall protection, 6.2, 6.6
hard hats, 6.3
working near water, 6.6
working with slings, 5.17
Personnel
for crane assembly, 3.5
crane operator, 3.4, 4.6, 4.14, 5.9, 5.25
foreman for unloading, 5.23
for hoisting operations
advanced, 6.8–6.9
crane and operational requirements, 6.5–6.7
lift plan, 6.15
overview, 6.1
trial lift, 6.1, 6.7–6.8
meetings, 4.15, 6.8
project engineer, 3.1
rigger. *See* Rigging specialist or superintendent
signal person, 4.6, 4.14, 4.15, 5.23, 5.25
site safety manager, 4.15
site superintendent, 3.1, 4.15
training, 4.25–4.26, 6.2
Pins. *See also* Fittings
A-frame, 2.7–2.8, 2.9, 2.10
attach, 2.8, 2.9
boom connector
in boom section installation, 2.12
in boom section removal, 2.13
in lattice boom assembly, 3.5, 3.6, 3.7, 3.8, 3.10
in lattice boom disassembly, 3.13
cable-retaining, 2.5, 2.6, 2.7, 2.9
cotter, 1.6, 2.13, 2.14, 2.15, 3.5
definition, 2.21
gear reducer lock, 2.15
hitch, 2.13, 2.14, 2.15
latch lock, 2.3, 2.4, 2.7, 2.11
lock, 2.14, 2.15
manual load, 2.13, 2.14, 2.15
offset adjustment, 2.12
pivot, 2.3, 2.6
retainer, 2.7, 2.9, 2.10, 2.13, 2.16, 6.3
stop-rail, 2.3
for swing-away, 2.5, 2.6, 2.12, 2.13
for telescope cylinder, 2.14
Pitting, 1.8
Plane, inclined, 5.16
Plastic, finish of wire rope, 1.4
Plate
data or identification, 3.6, 3.7, 6.3, 6.5
latch, 2.3, 2.4, 2.7, 2.11
Platform, personnel
boom-attached, 6.4–6.5
crane and operational requirements, 6.5–6.7
during crane travel, 6.7
inspection, 6.7, 6.8, 6.16
maximum load, 6.1, 6.3, 6.6
overview, 6.3
requirements, 6.2–6.5
trial lift, 6.7–6.8
Plow steel (PS), 1.2, 1.19
Plug
dummy, 2.13
pipe, 5.12
short-circuit, 2.13, 2.16
Pneumatic equipment, 6.5
Power, transmission through the crane, 1.4, 1.15
Power lines, 4.15, 6.8–6.9, 6.17–6.18
PPE. *See* Personal protection equipment
Pressure vessels, 5.1, 5.2
Prohibited zone, 6.8–6.9
PS. *See* Plow steel
Pullers, 5.11
Pump, hydraulic, 5.11, 5.12

Quadrants of operation, 5.6–5.9
Quality assurance / quality control, 4.14

Radio, 4.6, 4.15
Radius. *See* Load operating radius; Swing radius
Rails
grab or hand, 6.2, 6.3, 6.4
for heat exchanger removal, 5.32
side, on A-frame jib, 2.6
stop, on swing-away lattice extension, 2.3, 2.6
Ramps, 5.16

Rebar, 5.21, 5.23–5.24, 5.27
Reel, storage, 1.13, 1.14, 2.16
Reeving
  and capacity, 4.3
  definition, 1.20
  a drum, 1.13–1.14
  and load capacity, 1.11
  overview, 1.11–1.12
  for spooling rope, 2.16
  threading wire rope, 1.4
Resin, 1.4, 1.5
Responsibilities, 4.23–4.25
Rigging
  for critical lifts, 4.14, 4.22
  horizontal, 5.2
  in the lift plan, 4.2
  for personnel platform, 6.3–6.4
  for single-crane lifts, 4.2
  for up-ending and setting vertical vessels, 5.33–5.35
Rigging specialist or superintendent, 1.15, 2.16, 4.14, 4.23–4.24, 5.1, 5.25
Rollers
  for cable or hoist support, 2.11, 2.12, 4.5
  to move loads, 5.13–5.15
Roof sheeting, 5.21
Rope
  fiber, 5.23
  lanyard, 6.2
  rotation-resistant, 6.4, 6.5
  suspension, 3.5, 3.8
  wire. *See* Wire rope
  wire core manila, 6.2
Rotational forces or moments, 5.3–5.5

Saddle, of wire rope clip, 1.7
Safety
  during boom extension assembly, 2.2
  clearance for crane swing, 3.4
  during crane assembly, 2.1, 3.1
  during critical lifts, 4.14
  design factor for nominal breaking strength, 1.12–1.13, 1.19
  during jib assembly, 3.12
  during lattice boom assembly, 3.5, 3.6, 3.7
  manual lifting and carrying a load, 5.23
  monitoring, 6.2
  and personnel lifts, 6.6
  reference information for lift operations, 4.2
  use of rollers, 5.15
Safe working load (SWL), 5.15, 5.19, 5.20, 5.21
Sail area, 5.9
Scaffolding, 5.31
Scrubbing, 1.9
Sea conditions, 5.21
Seale strand pattern, 1.3, 1.4
Sections of a boom, intermediate, 2.11–2.13, 3.6–3.8, 3.9–3.11
Seizing, wire rope end preparation, 1.4, 1.10, 1.20
Shackle
  with multiple-crane lifts, 4.12, 4.13
  on personnel platform, 6.3, 6.4, 6.5, 6.13
  with slings, 5.17, 5.18
  with tailing lug, 5.33, 5.34
Shank, 1.5
Sheave(s)
  A-frame, 2.9, 2.10
  auxiliary boom-head, 2.15–2.16
  boom, 1.12
  boom-head, 2.5, 2.6, 2.12
  cable-roller, 2.5, 2.6, 2.9, 2.10
  definition, 1.10, 1.20
  diameter and D/d ratio, 1.10, 1.11
  extension, 2.5, 2.6
  and fleet angle, 1.14–1.15
  idler, 2.7, 2.10, 2.21
  inspection, 1.9–1.10
  load block, 1.12
  point, for spooling rope, 2.16
  and reeving configurations, 1.12
  spreading the wear on wire rope, 1.10
  threading wire rope, 1.4
Shipments, 5.23
Shutdown, 5.9
Signage
  identification plate for boom, 3.6, 3.7
  identification plate for personnel platform, 6.3, 6.5
  for power lines, 6.9
Signal person, 4.6, 4.14, 4.15, 5.23, 5.25
Signals, 5.24–5.25. *See also* Hand signals
Skids, 5.13, 5.15
Skiff, lifesaving, 6.6
Sleeve, swage, 6.3, 6.4
Sling(s)
  for boom, 3.3, 3.5
  bridle, 6.3, 6.4, 6.5, 6.13
  chain, 3.3, 3.5
  in the lift plan, 4.2
  to lift rebar bundles, 5.24
  measurement, 5.16
  selection and using, 5.16–5.17
  tension, 5.17–5.19, 5.24
  for vertical vessel up-ending and setting, 5.33–5.34
Sling angle, 5.17–5.18, 5.21, 5.33
Sling capacity chart, 5.18–5.19
Socket (electrical), 2.13, 2.16
Socket (mechanical)
  poured, 1.4–1.5
  swaged, 1.5
  wedge, 1.4, 1.5–1.6
Solenoid, 2.13, 2.15
Spelter, 1.4, 1.20
Splice, 1.4, 1.5, 1.9
Spool, 2.16
Spreader, hydraulic, 5.11
Stability
  of crane, 5.5–5.9, 5.10. *See also* Accidents, tipping; Tipping axis
  of jack system, 5.13
Steel
  in full-body harness, 6.3
  grades, 1.2, 1.4, 1.19
  in personnel platform, 6.2
Storage
  A-frame jib, 2.10–2.11
  boom, 2.6
  reel for wire rope, 1.13, 1.14, 2.16
  sling, 5.17
  swing-away lattice extension, 2.3, 2.6
  tools, 6.3
Strand patterns, 1.3, 1.4, 1.20
Strands, of wire rope
  configurations, 1.1, 1.2–1.4
  definition, 1.20
  inspection, 1.7, 1.8

Strength
D/d ratio and diameter of sheaves, 1.10, 1.11
nominal breaking
catalog, 1.12
definition, 1.19
for a wedge socket, 1.4
for wire rope, 1.12, 2.16
tensile, 1.8
Stress
bending and D/d ratio, 1.10, 1.11
load, on sling, 5.19, 5.20
and nominal breaking points, 1.4, 1.12
on sheaves, 1.10
on wire rope, 1.8, 1.9, 1.10, 1.20
Stretching, of wire rope, 1.7, 1.8
Strut, jib, 3.12
Submerged operations, 5.21, 5.24, 6.6
Suction, 5.24
Superstructure, 5.7
Swing-away, 2.3, 2.4, 2.5, 2.6, 2.12, 2.13
Swing-away lattice extension (fly), 2.2–2.6
Swing radius, 3.4
Switch(es)
anti-two-blocking, 2.12, 2.13
lifting limit, 2.6, 2.16
manual extension, 2.13, 2.15
Swivel, 1.5, 6.4
SWL. *See* Safe working load

Tail, of wire rope, 1.6
Tanks, 5.1, 5.2
Telescoping boom attachment. *See* Boom extension
Tests
lift operation practice, 4.15
of a load, 1.7
personnel lift, 6.1, 6.7–6.8
sling, 5.16
Thimble, 1.4, 1.5, 6.3, 6.4
Timbers, 2.1, 5.23
Tipping axis. *See also* Fulcrum
definition, 5.3, 5.27
principles of leverage, 5.3, 5.4, 5.5
relationship with center of gravity, 5.7–5.9, 5.24, 5.25
and submerged lifts, 5.24
Tires, 6.7
Toe board, 6.2, 6.3, 6.4
Torque
block twisting, 1.13
rotational force around a fulcrum, 5.4
in tightening nuts, 1.7
Traction steel (TS), 1.2
Training, 4.25–4.26, 6.2
Transport of crane, 3.1, 3.3
Treads, 3.1, 3.3, 3.9
Trucks, in the work area, 3.4, 5.23
Trunnion, 5.33
TS. *See* Traction steel
Tube bundle, 5.28–5.30
Tugger, 5.15
Turnbuckle, 2.9, 2.21

U.S. Coast Guard, 6.6

Valves, in jack system, 5.12
Vessels, vertical, 5.33–5.35
Vibration, of wire rope, 1.9, 1.10
Voice signals, 5.25

Warrington strand pattern, 1.3, 1.4
Water, lift operations over or near, 5.21, 5.24, 6.6
Wave action, 5.21, 5.24
Weather
and hoisting personnel, 6.6
as a potential hazard, 4.15
wind, 3.4, 5.9, 5.10
and working near power lines, 6.8
Weight
anti-two-blocking switch, 2.12
deductions, 4.3, 4.5
effects of buoyancy, 5.24
lifting limit switch, 2.6, 2.16
in the lift plan, 4.1, 4.31, 4.32
tare, 5.21
Weight deduction charts, 4.5
Welding
definition, 3.17
inspection of boom welds, 3.5
on personnel platforms, 6.3, 6.5
welded wire rope end preparation, 1.4
Wind, 3.4, 5.9, 5.10
Wire
definition, 1.20
filler, 1.3, 1.19
inner, 1.8, 1.19
preformed, 1.4, 1.19
Wire rope. *See also* Cable; Guide
broken, 1.7, 1.8, 1.9
components, 1.1–1.2
configuration, 1.2–1.4
dead end and live section, 1.6, 1.12, 1.19
diameter and sling capacity, 5.18
end terminations and preparations, 1.4–1.7, 1.8, 1.9
flat braided, 5.31
identification, 1.4
inspection, 1.7–1.9
installation, 1.13–1.14, 2.16
in lift plan, 4.31
maintenance, 1.10
parts of line for reeving, 1.12–1.13
reduction of diameter, 1.7, 1.8
replacement, 1.11
untwisting and the load block, 1.13
weight during lift, 4.2
Wire strand core (WSC), 1.2, 1.20
Work site
placement of crane on the, 3.4, 4.1
trucks on the, 3.4, 5.23
WSC. *See* Wire strand core

Yield point, 6.4, 6.13

Zinc, 1.4